Gas Chromatographic Headspace Analysis

Gas Chromatographic Headspace Analysis

Horst Hachenberg

and

Adolf P. Schmidt

Hoechst A.G., Frankfurt am Main, Federal Republic of Germany

Translated by

D. Verdin

HEYDEN

London · Philadelphia · Rheine

Heyden & Son Ltd., Spectrum House, Hillview Gardens, London NW4 2JQ, U.K.
Heyden & Son Inc., 247 South 41st Street, Philadelphia, Pennsylvania 19104, U.S.A.
Heyden & Son GmbH, Münsterstrasse 22, 4440 Rheine, West Germany.

Reprinted 1979

ISBN 0 85501 205 6

Printed in Great Britain by William Clowes & Sons Limited, Beccles and London

PREFACE

In the opinion of the authors many potential applications of the technique of gas chromatographic headspace analysis have yet to be fully exploited. This is true both for gas chromatographic trace analysis and for the measurement of physico-chemical data.

The aim of this book is to promote the wider use in research and industry of this simple and elegant measurement technique, which is economical in both time and effort.

The first part, by H. Hachenberg, is concerned mainly with applications of the technique to the trace analysis of materials for which normal gas chromatographic analysis is very complicated or even impossible to perform. The majority of the examples which have been discussed in detail are those which have already been developed into quantitative analytical procedures.

In the second part of the book A. P. Schmidt explains how the theoretical basis of quantitative headspace analysis may be applied to the measurement of a wide range of physico-chemical data.

The book is written for analysts working in the fields of chemistry and medicine, and equally for physical chemists and industrial technologists.

The authors thank Hoechst AG for permission to publish this book, and Perkin-Elmer and Co. GmbH, Bodenseewerk, Überlingen, Germany, for providing some of the illustrations.

H. Hachenberg
A. P. Schmidt

Frankfurt am Main,
October, 1976.

CONTENTS

PART 2

Physico-Chemical Applications

SYMBOLS

1. Capital letters

A^*	constant in the Antoine vapour pressure equation for a given compound
B^*	constant in the Antoine vapour pressure equation for a given compound
B	second virial coefficient
B_{ij}	second virial coefficient for the interaction between i and j
C^*	constant in the Antoine vapour pressure equation for a given compound
C_i	concentration of i in the liquid phase
C'_i	equilibrium concentration of i in the liquid phase
C''_i	equilibrium concentration of i in the vapour phase
E_T	transition energy (solvent parameter of Dimroth and Reichardt)
E_{HS}	sensitivity of the gas chromatographic method of headspace analysis
F_i	peak area measured for substance i in gas chromatographic headspace analysis of a liquid mixture containing i
F_{0i}	peak area obtained for i in the gas chromatographic headspace analysis of pure liquid i
F''_i	peak area obtained for i in the gas chromatographic analysis of a vapour sample
F'_i	peak area obtained for i in the gas chromatographic analysis of a liquid sample
$\bar{F}_i$	normalized peak area used to calculate the amount of i in gas chromatographic headspace analysis
G_i	partial molar free enthalpy of i
G_{0i}	standard molar free enthalpy of i
G_i^E	partial molar excess free enthalpy of i
$\bar{G}^E$	mean molar excess free enthalpy
H_i	partial molar enthalpy of i
H_{0i}	standard molar enthalpy of i
$\bar{H}_i^E$	partial molar excess enthalpy of i
$\bar{H}^E$	mean molar excess enthalpy
K_i	partition coefficient of i in gas chromatography
L	solvent

L_p	molar heat of vaporization
M_L	molecular weight of L
P	total vapour pressure
P_c	critical pressure
R	molar gas constant
$\bar{S}^E$	mean molar excess entropy
S	selectivity
S^*	selectivity at infinite dilution
T	absolute temperature
T_c	critical temperature (absolute)
V	volume
V'	volume of the liquid phase
V''	volume of the vapour phase
V_i	partial molar volume of i
V_{0i}	molar volume of i
V_i^E	partial molar excess volume of i
$\bar{V}$	mean molar volume
$\bar{V}^E$	mean molar excess volume
V_g^T	specific retention volume corrected for the pressure drop of the column at temperature T

2. Lower case letters

a_i	activity of i
$\bar{c}_i$	calibration factor for a given substance i with a particular detector in gas chromatographic analysis
c_i	calibration factor for a given substance i with a particular detector in gas chromatographic analysis
d_L	density of L
f_i	fugacity of i
h_i	peak height for i
i	substance i
j	substance j
m_i	mass of i
n_i	number of moles of i
n'_i	number of moles of i in the liquid phase
n''_i	number of moles of i in the vapour phase
p	vapour pressure of a pure substance
p_i	partial pressure of i
p_{0i}	vapour pressure of pure i
r_i	calibration factor for a given substance i with a particular detector in gas chromatographic analysis
x_i	mole fraction of i in the liquid phase
y_i	mole fraction of i in the vapour phase
t_i	apparent retention time of i

3. Greek letters

α_{12}	relative volatility of substances 1 and 2
$(\alpha_{12})_{123}$	relative volatility of substances 1 and 2 in a mixture of three substances 1–2–3
$(\alpha_{12})_{12\infty3}$	relative volatility of substances 1 and 2 at infinite dilution in substance 3
β	improvement factor
β^*	improvement factor at infinite dilution
γ_i	activity coefficient of i
$\gamma_{i\infty L}$	activity coefficient of i at infinite dilution of i in L
δ	Hildebrand's solubility parameter
ε^0	solvent strength of Snyder
φ_i	fugacity coefficient
ϕ_i	real gas correction factor for substance i
ϑ	temperature/°C

INTRODUCTION

Gas chromatographic headspace analysis is an indirect method for the determination of volatile constituents in liquids or solids by gas chromatographic analysis of the vapour phase which is in thermodynamic equilibrium with the sample to be analysed in a closed system. It is therefore a specialized form of gas analysis and does not represent a fundamentally new analytical method, since appropriate techniques—for example, mass spectrometry and IR and UV spectroscopy—already exist for the determination of compounds in their vapour state in the presence of liquids and solids. However, these methods are often insufficiently sensitive, and when several materials are present in the vapour phase, they give inextricably complex results, since only the sum of the components can be measured.

Headspace analysis has achieved considerable importance as a consequence of its use in conjunction with gas chromatography, where the efficient resolution available and the high sensitivity and selectivity of gas chromatographic detectors can all be exploited. It is particularly suited to the determination of trace components in samples which cannot be handled by the normal gas chromatographic techniques because decomposition occurs when the sample is vaporized, or because dissociation products not originally present in the material are formed during the sampling procedure. The applications of headspace analysis include the determination of solvent residues in solid chemicals and packaging materials, the residual monomers and olfactory substances in plastics and polymer dispersions, and the analysis of aromatic ingredients in basic and luxury foods. They also include problems involving the trace analysis of aqueous materials which are difficult to carry out by conventional gas chromatography such as, for example, drinking-water, sewage, alcoholic and non-alcoholic beverages, and, especially, the analysis of blood alcohol, which has provided an impetus for the actual development of this method of analysis.

Gas chromatographic headspace analysis is therefore a specific form of gas chromatographic trace analysis[1,2] and is employed almost exclusively in this sense. The advantages of this method of headspace analysis compared with the conventional manipulation of samples requiring lengthy preliminary procedures such as extraction, distillation, steam distillation, etc., for the concentration of the trace constituents are quite obvious. They result from the economy of effort and from the fact that no overloading or contamination of the column with water and high-boiling or non-volatile materials occurs, such as can happen with a direct sampling technique. Since low concentrations of the components already exist in the vapour phase, the sample does not have to be vaporized for the gas chromatographic analysis and, consequently, better chromatographic separations are generally achieved than in conventional gas chromatography.

Gas chromatographic headspace analysis has almost become the established practice for the investigation of problems involving odours, and is therefore occasionally also described as the 'gas chromatographic nose', although this is rather inaccurate in that the specificity of the human nose is not always the same as that of the gas chromatographic detector employed.

In contrast to conventional gas chromatography, problems arise both in the experimental technique and in the qualitative and quantitative interpretation of results. They arise from the fact that this elegant and powerful analytical method is not yet used to an extent consistent with its importance. Thus, qualitative gas chromatographic headspace analysis of unknown samples is still less positive than classical gas chromatography, since the concentrations in the vapour phase are too low for trace components to be isolated to enable them to be identified. Whereas this problem can often be solved in classical chromatography by increasing the sample size, in the case of the gas chromatographic headspace analysis of unknown samples it is only possible to record comparative chromatograms, i.e. 'fingerprints', unless a combined gas chromatograph–mass spectrometer is available. The difficulties in quantitative evaluation are due to the fact that the concentrations of the components being determined in the sample have to be derived from the measurement of their concentrations in the headspace. Compared with classical gas chromatography, this gives rise to additional parameters which have to be considered, but which, on the other hand, open up an interesting and wide field of non-analytical applications of gas chromatographic headspace analysis, namely the measurement of thermodynamic data.

PART 1

Analytical Applications

1.1. THE THEORETICAL BASIS OF QUANTITATIVE ANALYSIS

The parameter measured in gas chromatographic headspace analysis is the peak area F_i, which, at relatively low pressures in the vapour phase above the sample, is proportional to the partial vapour pressure p_i of the volatile component i in the headspace:

$$F_i \sim p_i \tag{1}$$

In the equation

$$F_i = \bar{c}_i p_i \tag{2}$$

derived from the above relationship, for a given sample size each substance has a specific value for the constant $\bar{c}_i$ which depends on the characteristics of the detector being used.

The partial vapour pressure p_i is generally expressed as

$$p_i = p_{0i} x_i \gamma_i \quad \text{(real system)} \tag{3}$$

where p_{0i} is the vapour pressure of pure component i, x_i is the mole fraction of component i dissolved in the sample and γ_i is the activity coefficient of the dissolved component i.

If, as only happens in rare cases, the liquid sample consists of an ideal mixture, then the activity coefficient γ_i has the value of 1 and Eq. (3) takes the form of Raoult's law:

$$p_i = p_{0i} x_i \tag{3a}$$

Examples of such mixtures are: *n*-heptane/methyl cyclohexane, *p*-xylene/*m*-xylene, 2-methylnaphthalene/1-methylnaphthalene, *trans*-decalin/*cis*-decalin, *o*-nitrochlorobenzene/*p*-nitrochlorobenzene, hexane/heptane, benzene/toluene, benzene/1,2-dichloroethane, chlorobenzene/ethylbenzene, ethanol/isopropanol.

However, in most cases the samples consist of non-ideal mixtures having partial vapour pressures which either show positive ($\gamma_i > 1$) or negative ($\gamma_i < 1$) deviations from Raoult's law.

Examples of non-ideal mixtures of the first type (positive deviations from Raoult's law) are: heptane/toluene, octane/ethylbenzene, methanol/isoamylalcohol, ethanol/isoamylalcohol, cyclohexane/dioxane, diethyl ether/ethanol,

ethanol/methylethylketone, ethanol/ethyl acetate, ethanol/heptane, benzene/aniline, toluene/phenol, dichloroethane/propanol, ethanol/water, ethanol/acetonitrile.

Examples of non-ideal mixtures of the second type (negative deviations from Raoult's law) are: acetone/1,1,2-trichloroethane, *trans*-dichloroethylene/di-isopropyl ether, *trans*-dichloroethylene/tetrahydrofuran, *cis*-dichloroethylene/methylethylketone, formaldehyde dimethylacetal/*trans*-dichloroethylene, methyl acetate/*cis*-dichloroethylene, acetone/*cis*-dichloroethylene, methanol/dimethylformamide, chloroform/methylisobutylketone, acetone/chloroform.

The activity coefficient γ_i depends not only on the nature of the component i and the other components in the mixture, but also on the mole fractions of all of the components. Moreover, it depends on the temperature and to a small extent on the pressure. The variation of the activity coefficients with concentration in a mixture at a given temperature and pressure is described by the Gibbs–Duhem equation.[3]

The interpretation of activity coefficients usually consists of an analysis of the intermolecular attractive forces (London-type forces, dipolar interactions, forces due to hydrogen bonding, and specific interactions) between like or unlike nearest-neighbour molecules in a liquid mixture, and these interactions can be of comparable magnitude (ideal systems) or dissimilar (real systems). For a complete understanding it is often necessary to include consideration of dissociation or association phenomena.

Although it rarely needs to be considered in practical gas chromatographic headspace analysis, if the original assumption of a low pressure in the vapour phase above the sample does not hold ($P \geqslant 2$ bar), then it is necessary to use more accurate vapour–liquid equilibrium relationships, which are derived by introducing fugacities,[3] and so contain a correction factor for real gases (see Part 2).

The quantitative interpretation of gas chromatographic headspace analysis depends on the equilibrium expression derived from Eqs. (2) and (3):

$$x_i = \frac{F_i}{\bar{c}_i p_{0i} \gamma_i} \qquad (4)$$

where $1/(\bar{c}_i p_{0i} \gamma_i)$ is a constant (calibration factor) which has to be determined by measurement.

1.2. THE CALIBRATION OF QUANTITATIVE ANALYSES

It is clear from Eq. (4) that the activity coefficient plays an important role in gas chromatographic headspace analysis. If the product $\bar{c}_i p_{0i} \gamma_i$ is known, then the concentration being determined can be calculated directly from the peak area obtained by gas chromatography. However, in almost all practical situations this product must be evaluated from measurements. For this purpose three cases can be distinguished:[4]

$$\begin{array}{ll} \gamma = 1 & \text{(ideal mixture)} \\ \gamma = \text{const.} & \text{(ideal dilute solution)} \\ \gamma = f(x) & \text{(real mixture)} \end{array}$$

An example of the calibration of an ideal mixture over the whole range of concentrations is shown by ethyl acetate/vinyl acetate (Fig. 1).

Since the practical application of gas chromatographic headspace analysis is almost exclusively for trace analysis, i.e. the determination of low concentrations, it normally involves the use of ideal dilute mixtures. Under these conditions the activity coefficient is constant, and therefore the product $\bar{c}_i p_{0i} \gamma_i$ is also constant if the detector employed exhibits a linear or proportional response ($\bar{c}_i$ = const.).

An example of a calibration plot for an ideal dilute mixture is shown by low concentrations of acrylonitrile in water[5] (Fig. 2).

If it is necessary to deal with samples in which relatively high concentrations of i are present, so that $\gamma = f(x_i)$ and the partial pressure and the peak area are also functions of the concentration, then a calibration curve must be plotted with the use of standard solutions of various concentrations. A system of this type is the well-known case of a binary mixture of ethanol and heptane.

Figure 3 shows the peak area obtained from gas chromatographic headspace analyses of Kolb[4] plotted as a function of the ratio (mole fractions) of ethanol to heptane. The results show that with increasing concentration the peak areas become non-linear, although the calibration plots can still be used as long as they are not horizontal. It is therefore simpler in practice to dilute the sample concerned to such an extent that the calibration can be performed in the region where the diluted solution exhibits ideal behaviour. The actual performance of the calibration and of the quantitative analysis therefore often requires significantly more effort than with conventional gas chromatography.

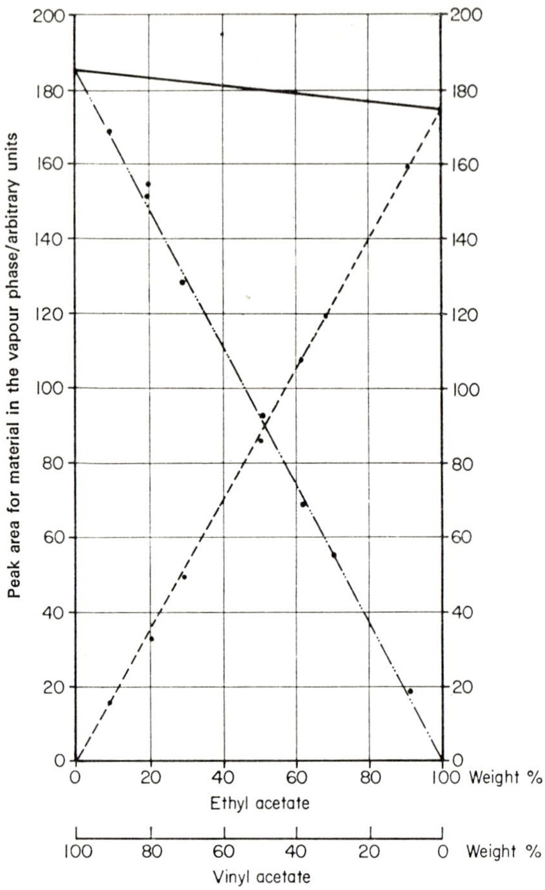

Fig. 1 Headspace analysis of ethyl acetate/vinyl acetate mixtures at 70°C.

Since at a given temperature the vapour pressure p_{0i} is characteristic of a compound, then the calibration must be performed under the same experimental conditions as those under which the analysis is to be carried out.

Because of the temperature dependence of the vapour pressure in accordance with the Clausius–Clapeyron equation

$$\frac{d \ln p}{dT} = \frac{L}{RT^2} \quad (L = \text{latent heat of vaporization}) \tag{5}$$

it is necessary to precisely control the temperature for the analysis. It is also important to know the time required to establish the vapour pressure equilibrium, since under certain circumstances it can be very dependent on the viscosity of the sample.

On the other hand, the activity coefficient is not purely a characteristic of a

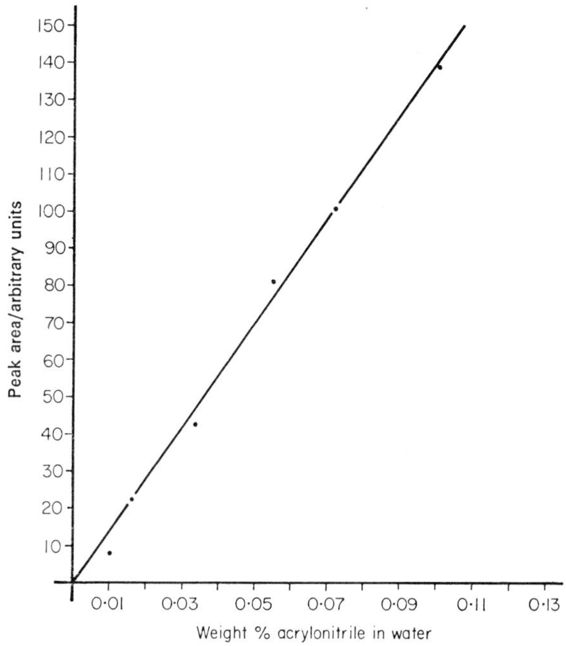

Fig. 2 Calibration of low concentrations of acrylonitrile in water.

compound. The value of γ_i depends on the composition of the mixture in which the substance i is present. The calibration must therefore be performed with a mixture the composition of which corresponds to that of the sample. As a consequence of this, considerable difficulties often arise with unknown samples. Thus, for example, the analysis of sewage can be erroneous if the calibration is carried out in pure water, since the presence of unknown concentrations of salts or other components can have a considerable effect on the activity coefficients. This problem may be overcome in the following ways.[4]

(1) The composition of the sample is known and it can be reproduced or simulated for the purposes of calibration. Thus, for example, for the determination of amyl alcohols in spirits calibration in aqueous ethanol solution is satisfactory.

(2) The composition of the sample is not known but it is available in a pure form. Examples of this are the determination of vinyl chloride in olive oil and of ethyl alcohol in blood.

(3) The composition of the sample is unknown and it is not available in a pure form. This happens, for example, in the analysis of sewage or in the determination of residual monomers in polymers of unknown origin. In this case the method of mixtures or a peak enhancement technique must be used.

Fig. 3 Headspace analysis of ethanol/heptane mixtures at 70°C (x = mole fraction).

The latter consists of analysing the sample under absolutely identical conditions (sample size, temperature of the vapour phase, gas chromatographic parameters) with and without the addition of the pure constituent being determined. From the increase in the peak area for the sample to which material has been added the quantity of the desired constituent in the original sample can be calculated. However, since in the case of unknown samples it is never quite certain whether the correspondence of the retention times in the samples establishes that the additive and the unknown are the same, then the result of the analysis remains open to question.

Every application of gas chromatographic headspace analysis involves calibration by an absolute method, since the vapour pressure p_{0i} is a characteristic property of a compound. The following expression may be used for this purpose:

$$C_i = \frac{F_i \cdot C^*_i}{F^*_i} \tag{6}$$

where F_i is the peak area for i in the analysis; C^*_i is the concentration of i in the standard sample; and F^*_i is the peak area for i in the standard sample. It is assumed that for both the calibration and the analysis the experimental conditions will be absolutely identical, and the medium will have the same composition. It is clearly only possible to perform measurements of this type with equipment designed specially for gas chromatographic headspace analysis.

Small changes in the chromatographic conditions (change in C_i) and in the composition or the equilibrium temperature of the sample (changes in the product $\gamma_i p_{0i}$) may be compensated for by using an internal standard, but it must be of approximately the same polarity as the substance being determined, so that the changes in the activity coefficients of both components are of comparable magnitude. Since in this case the internal standard not only compensates for the detector response, as in conventional gas chromatography, but also simultaneously takes account of variations in the medium and the temperature, Kolb[4] has suggested the term 'internal compensator'.

The value of the calibration factor is obtained from

$$\bar{c}_i = \frac{F_{st} \cdot C_i}{C_{st} \cdot F_i} \tag{7}$$

and the result of the analysis is calculated from

$$C_i = \frac{F_i \cdot \bar{c}_i \cdot C_{st}}{F_{st}} \tag{8}$$

in which $\bar{c}_i$ is the calibration factor; F_{st} is the peak area for the internal standard; C_i is the concentration of i in the sample; C_{st} is the concentration of the internal standard; and F_i is the peak area for i.

In gas chromatographic headspace analysis the concentration C is usually expressed in units of p.p.m. by weight with respect to the total weight of the sample. If an integrator or computer is not available, calculation in terms of the peak height is satisfactory, since at the low concentrations involved the peak height is proportional to the amount of the component present.

1.3. METHODS OF INCREASING THE ANALYTICAL SENSITIVITY

For the quantitative determination of known trace components it is sometimes necessary to increase the sensitivity towards materials in the vapour phase. The normal method used in classical gas chromatography of increasing the volume of the sample is not very effective in this case, since the concentrations in the headspace are usually low. It is therefore recommended that specific detectors be employed for the analysis of particular groups of compounds, e.g. the flame photometric detector for sulfur compounds and the electron capture detector for halogenated compounds or diketones.

This, and other techniques used in gas chromatographic trace analysis, are well known[1] and will therefore not be discussed any further here. On the other hand, it can be inferred from Eq. (4) that some additional techniques are possible in the case of gas chromatographic headspace trace analysis. The sensitivity of this method is often called the Headspace Response, and is related by the following equation to the concentration x_i of the trace component in the sample being analysed:

$$E_{HS} = \frac{F_i}{x_i} = \bar{c}_i p_{0i} \gamma_i \qquad (9)$$

It can be increased by the methods described below.

(1) Increasing the sensitivity by raising the temperature. The simplest method of enhancing the sensitivity E_{HS} is to increase the saturated vapour pressure p_{0i} of the pure trace component i. Equation (5) shows that this can be achieved by raising the temperature. However, this technique is of limited experimental application, owing to the risk of bursting the container or of losing the components as a result of chemical interaction with the material from which the septum used to seal the vessel is made.

(2) Increasing the sensitivity by increasing the value of γ_i. The sensitivity may also be improved by increasing the activity coefficients γ_i. This can be achieved by adding to the sample a compound which can be either (a) an electrolyte or (b) a non-electrolyte.

(a) The addition of electrolytes to increase γ_i. The 'salt effect'—that is, the effect of the ions of dissolved salts on the activities of substances dissolved in aqueous solutions—has long-standing applications in analytical and preparative chemistry as a means of enhancing the solubility of certain dissolved substances ('solubilization'), or of lowering their solubility ('salting out'). This effect only occurs when the ions in an aqueous medium have a substantial influence on the activities and the corresponding activity coefficients, and can be interpreted in terms of particularly strong intermolecular electrostatic interactions. The

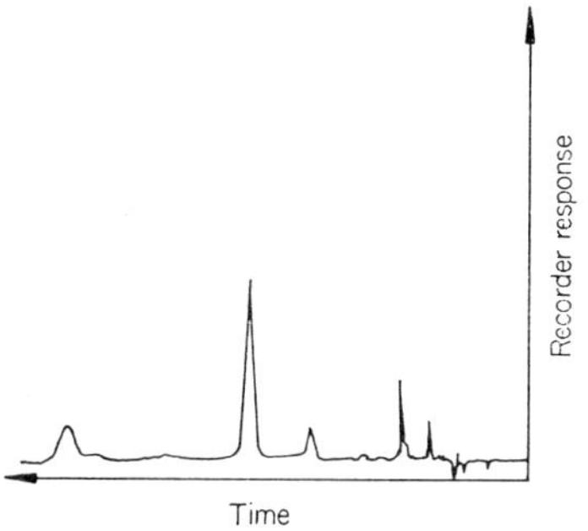

Fig. 4 Gas chromatographic headspace chromatogram of sewage containing 30% added K_2CO_3.

'salting-out' effect[6,7] is very often employed for aqueous samples as a means of increasing the concentration of volatile trace components being analysed in the headspace. As an example, Fig. 4 shows the gas chromatographic headspace analysis of sewage to which about 30% K_2CO_3 has been added, and it may be compared with Fig. 5 for the analysis of the original sample without any added salt.

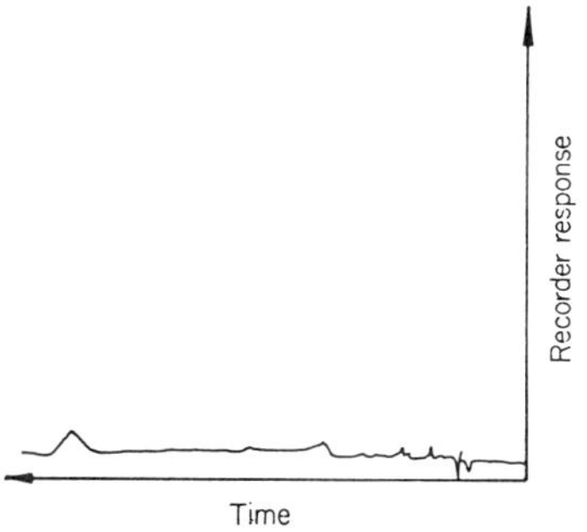

Fig. 5 Gas chromatographic headspace chromatogram of sewage with no added salt.

The magnitude of this salting-out effect often depends on the nature of the salt employed, as shown by the following example of a 2% solution of ethanol in water.[8] If 0.5 g of various anhydrous salts is added to this solution, the concentration in the headspace increases relative to that above the original solution by the factors shown in Table 1.

TABLE 1

Increases in the concentration of ethanol in the headspace due to the addition of different salts (temperature 60°C)

Salt	*Increase in concentration in headspace*
Ammonium sulfate	×5
Sodium chloride	×3
Potassium carbonate	×8
Ammonium chloride	×2
Sodium citrate	×5

In this connection it should be noted that in studies of the effect of temperature on the vapour pressure equilibria in systems containing such salt additives it has been observed that under certain circumstances the Clausius–Clapeyron equation is not obeyed exactly.[9]

(b) The addition of non-electrolytes to increase γ_i. The analytical sensitivity may also be enhanced by increasing the activity coefficients of dissolved compounds by adding non-electrolytes to the sample being analysed. For example, the sensitivity towards organic materials in organic solvents which are miscible with water may be improved considerably by the addition of water. An example which illustrates this is the addition of increasing amounts of water to a solution containing 120 p.p.m. by weight of styrene, butyl acrylate, *n*-butanol or acrylonitrile in dimethylformamide (DMF). The resulting increases in the peak area for the vapour phase are shown in Table 2 and Fig. 6.[10]

TABLE 2

The peak area due to 120 p.p.m. by weight of styrene, butyl acrylate, *n*-butanol and acrylonitrile in DMF as a function of the water content of the mixture

Mixture	*Peak area from headspace analysis*			
H_2O:*DMF*	*acrylonitrile*	n-*butanol*	*butyl acrylate*	*styrene*
0:100	12	2	3	4
10: 90	18	3	7	9
20: 80	25	5	15	21
30: 70	45	9	51	81
40: 60	58	14	83	144
50: 50	71	18	122	227
60: 40	87	23	179	344
70: 30	105	30	243	458
80: 20	118	37	280	556
90: 10	119	45	307	504
100: 0	139	51	334	600

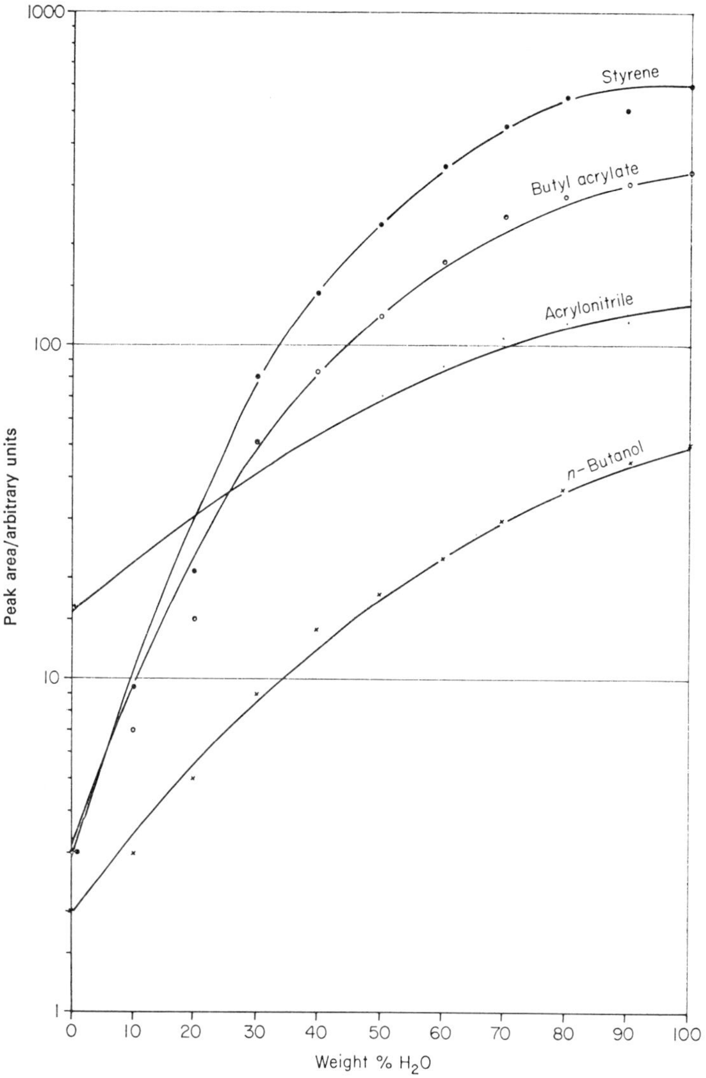

Fig. 6 The peak area due to 120 p.p.m. by weight of styrene, butyl acrylate, *n*-butanol and acrylonitrile in DMF as a function of the water content of the mixture.

The results in Table 2 and Fig. 6 clearly demonstrate the effect on the sensitivity of this gas chromatographic headspace analysis caused by the addition of water. As is to be expected, the effect is greatest with styrene and butanol shows the smallest change. In this connection, interesting possibilities arise in an homologous series, as illustrated by a mixture of the C_2–C_5 n-alcohols in DMF. Water is again used as the additive and the experimental results are presented in Table 3 and Fig. 7.

TABLE 3

Peak area from the headspace analysis of 120 p.p.m. by weight of each of the C_2–C_5 *n*-alcohols in DMF as a function of the water content of the mixture

Mixture H_2O:*DMF*	*ethanol*	n-*propanol*	n-*butanol*	n-*pentanol*
	*Peak area**			
0:100	12.1	4.9	2.1	0.9
20: 80	13.3	7.3	4.2	2.4
30: 70	15.0	9.4	6.2	3.6
40: 60	15.7	11.4	8.5	5.6
60: 40	18.9	16.6	14.9	14.2
70: 30	22.1	21.4	21.9	23.2
80: 20	22.1	23.3	25.7	29.7
100: 0	26.0	32.5	44.5	63.2

* *Corrected for variations in the detector response.*

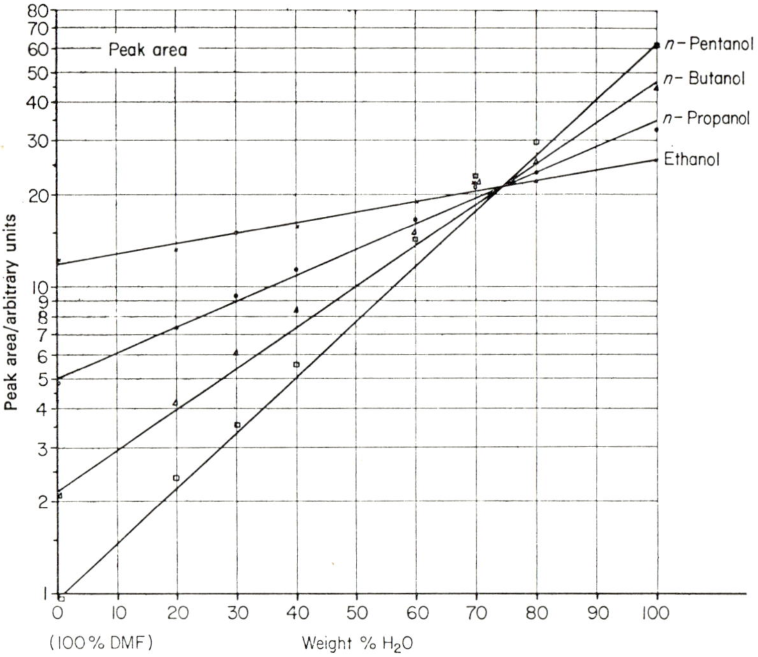

Fig. 7 Peak area from the headspace chromatograms for 120 p.p.m. by weight of each of the C_2–C_5 *n*-alcohols in DMF as a function of the water content of the mixtures.

It can be inferred from Fig. 7 that the sensitivity of the gas chromatographic headspace analysis of alcohols dissolved in DMF is increased by the addition of water. The sensitivity itself and the increase in it display systematic differences through the homologous series. The sensitivity towards the homologues in DMF solution diminishes as the length of the carbon chain increases, but the increase in the sensitivity on adding water increases with the chain length. Since the rate of rise remains constant over the whole range of DMF–water mixtures, there is an inversion in the order of sensitivities in aqueous solutions, so that the sensitivity then increases with the length of the carbon chain. There is a conspicuous point at about 70–80 weight % water characterized by the fact that it represents the sensitivity of the gas chromatographic headspace analysis for all of the homologous alcohols which have been studied. Obviously at this point the activity coefficients, which depend on the concentration of water and which can be used to explain quantitatively the influence of the water content on the sensitivity of gas chromatographic headspace analysis, are changing systematically in such a way that they completely compensate for the accompanying

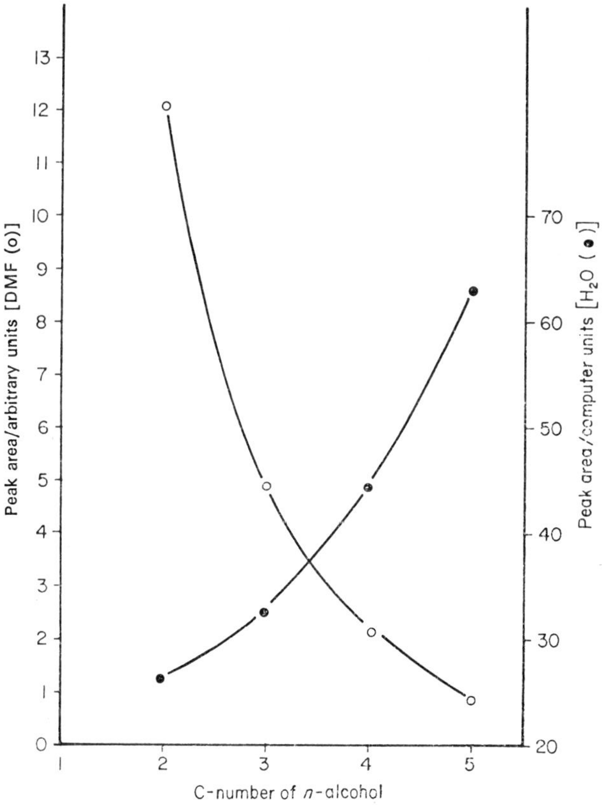

Fig. 8 Peak areas for C_2–C_5 *n*-alcohols in pure DMF and in pure water as a function of the C-number.

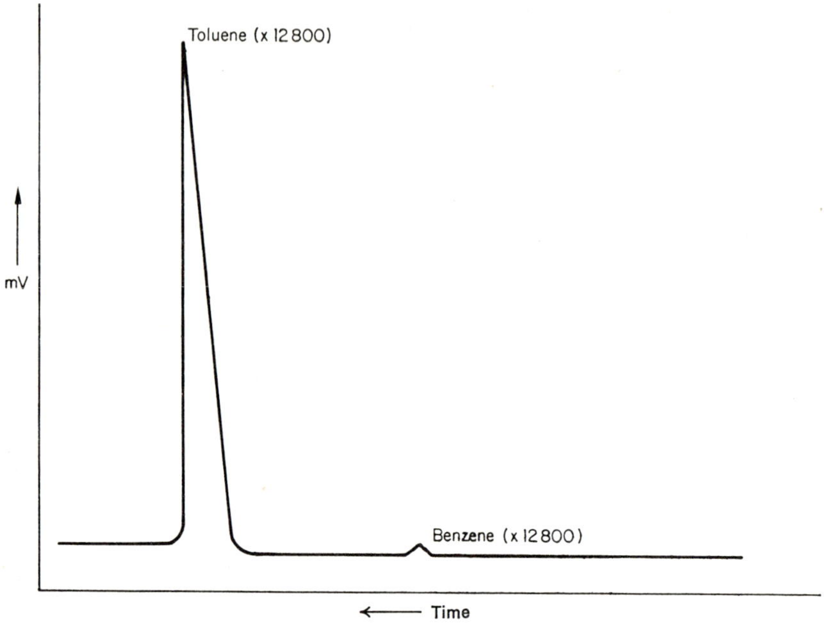

Fig. 9 Gas chromatogram of a mixture of 99 vol.% toluene and 1 vol.% benzene produced by sampling the liquid mixture.

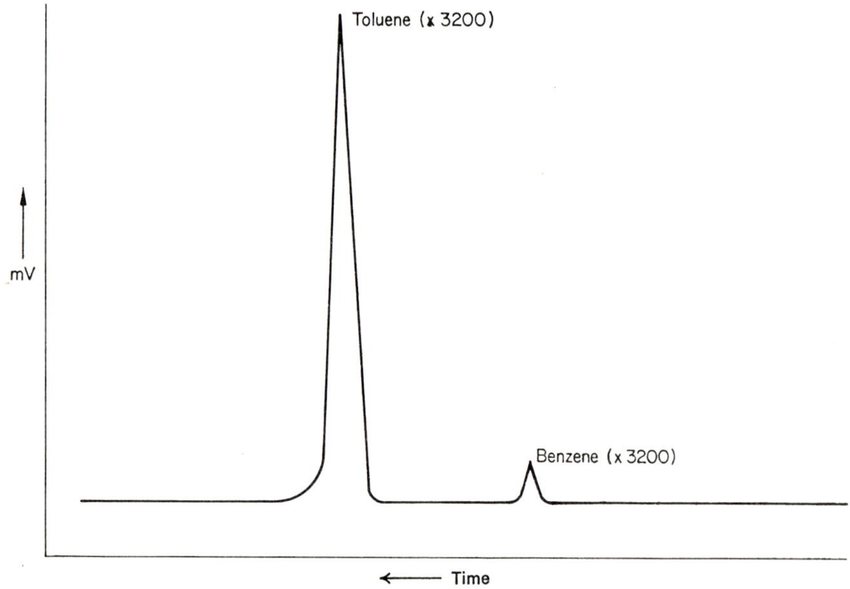

Fig. 10 Gas chromatogram of a sample from the vapour phase above a mixture of 99 vol.% toluene and 1 vol.% benzene.

variations in the saturated vapour pressures p_{0i} of the pure homologues, with the result that the product $p_{0i}\gamma_i$ is constant for the series of homologues investigated. Conversely, the sensitivity of ethanol can, of course, be enhanced relative to that of pentanol by adding DMF to their aqueous solutions. The greater the differences in molecular structure and polarity the greater the difference in sensitivity, as is clearly demonstrated for the C_2–C_5 *n*-alcohols in Fig. 8, which is based on data taken from Table 3.

As a consequence of the fact that the activity coefficient plays a decisive role in determining the sensitivity, many gas chromatographic trace analyses are better performed as headspace analyses than in the liquid phase. This is always the case when the concentration of the trace component being determined is enhanced relative to that of the main component or other trace components in the headspace, i.e. when the relative volatility

$$\alpha_{12} = \frac{p_{01}}{p_{02}}\left(\frac{\gamma_1}{\gamma_2}\right) \tag{10}$$

of the trace component relative to that of the main component or another trace component is $\gg 1$. To achieve the largest possible effect the temperature at which the liquid sample is held must be so chosen that the maximum difference exists between the vapour pressures.

Thus, for example, in the determination of traces of benzene in toluene Jentzsch and co-workers[11] have found that the ratio of the peak areas for benzene and toluene in the headspace is greater than that in the liquid phase. Toluene containing 1 vol. % benzene was analysed by conventional sampling of the liquid mixture and also by sampling the headspace above it. The chromatogram obtained for the sample taken from the liquid phase is shown in Fig. 9 and that for the headspace sample is shown in Fig. 10. Whereas the composition of the sample in the liquid phase leads to a ratio of 1:100 for the two components,

TABLE 4

Comparison of the relative peak areas (F_{rel}) (acetone = 1.00) obtained for 100 p.p.m. by weight of organic compounds in water by sampling of the liquid phase and of the headspace

Compound	*Liquid phase sampling**	*Headspace sampling†*	$\frac{F_{rel}\ \text{headspace}}{F_{rel}\ \text{liquid}}$
Acetaldehyde	0.37	0.5	1.35
Methyl acetate	0.58	1.8	3.1
Ethanol	0.1	0.15	0.19
Ether	1.42	13.2	9.3
Pentane	1.8	3.5	1.9
Benzene	2.3	35.5	15.4
Acetone	1.0	1.0	1.0

* *Sample size, 2 μl.*
† *Model F40 chromatograph; sampling time, 6 s.*

a ratio of 4.3:100 is deduced from the chromatogram obtained by headspace analysis, i.e. the benzene concentration has increased by a factor of 4.3 in the sample injected into the gas chromatograph.

Another example of this is the determination of 100 p.p.m. by weight of acetaldehyde, methyl acetate, methanol, diethyl ether, *n*-pentane, benzene and acetone in water.[12] The relative peak areas (acetone = 1.00) obtained by sampling of the liquid are compared with those measured for the headspace in Table 4.

It follows from Table 4 that the sensitivity of gas chromatographic headspace analysis is higher than that achieved with sampling of the liquid phase by the following factors relative to acetone = 1.0: benzene 15.4, ether 9.3, methyl acetate 3.1 and *n*-pentane 1.9. On the other hand, the sensitivity towards ethanol is 0.19 times that of acetone.

This technique does not, of course, increase the absolute detectable amount of a component i in the liquid sample, since it can only give rise to a relative increase in the sensitivity of that component compared with another component in the vapour phase. However, it is to be recommended in many cases of trace analysis involving poor chromatographic resolution.

1.4. POSSIBLE APPLICATIONS OF QUALITATIVE GAS CHROMATOGRAPHIC HEADSPACE ANALYSIS

Disregarding the identification of materials on the basis of measurements of retention data, which is extremely unreliable even in the classical gas chromatographic analysis of unknown samples, the possible applications of quantitative headspace analysis are very limited, owing to the exceedingly low concentrations in the vapour phase. The normal enrichment techniques used in chromatographic gas analysis[13] cannot be fully utilized, since they require large volumes of vapour which do not exist in the case of headspace analysis. The few methods of this kind which can be used are based mainly on microcondensation or chemical sorption techniques, and will be dealt with in the discussion of sampling.

Alternative approaches consist of the established methods of identifying types of compounds by using detectors which are specific to certain substances,[14] or by means of liquid phase chemical reactions of the headspace vapours.[15] Thus, ketones can be removed from the headspace with acidified hydroxylamine solution, sulfur compounds with mercuric chloride, and esters with alkaline hydroxylamine solution. Palo described the use of this identification technique in the determination of fundamental groups in the volatile compounds from dairy products.[16]

The qualitative gas chromatographic headspace analysis of unknown samples can only really be carried out in association with a mass spectrometer, and for the reasons already mentioned only very high sensitivity mass spectrometers are suitable.[17–20]

1.5. THE PRACTICAL ASPECTS OF GAS CHROMATOGRAPHIC HEADSPACE ANALYSIS

1.51. General

The gas chromatographic analysis of vapours which are in equilibrium with a liquid phase in a closed system requires a special experimental technique quite different from that used in ordinary gas analysis. The instrumental parameters must be such as to maintain the sample container at a constant and reproducible temperature. This also applies to the handling of the specimen during its preparation and during the actual sampling from the headspace vessel, since condensation and fractionation of components can easily occur at this stage and affect the integrity of the qualitative analysis and the accuracy of the quantitative determination. Specially designed equipment is therefore essential for accurate gas chromatographic headspace analysis.

1.52. The Headspace Analysis Vessel

In principle any sealed container having a point from which material in the vapour phase can be withdrawn is suitable as a vessel for headspace analysis. The simplest and also the most frequently used arrangement is shown in Fig. 11.

In addition to this experimental procedure in which the sample has to be taken by means of a syringe, the well-known method of partial pressure sampling used in gas analysis can also be employed[21] (see Fig. 12). In this method the use of a reduced pressure enables a definite amount of material to be reproducibly removed from the headspace and introduced directly into the gas sampling valve on the gas chromatograph. It is important that the gas sampling valve and its associated tubing be heated, so that condensation cannot occur.

Attention should also be paid to the following points when a headspace analysis vessel is selected:

(1) If a rubber septum is employed, the exposed surfaces should be kept as small as possible, to minimize effects due to dissolution and diffusion. In addition, the internal surface of the septum can be covered with a metal foil.

(2) The volume of the headspace vessel should not be too large compared with that of the sample, since under certain circumstances it might result in total evaporation of the component being analysed into the gas phase. The assumptions made for quantitative analysis would no longer be satisfied, since the sample would not then be in equilibrium with the vapour phase.

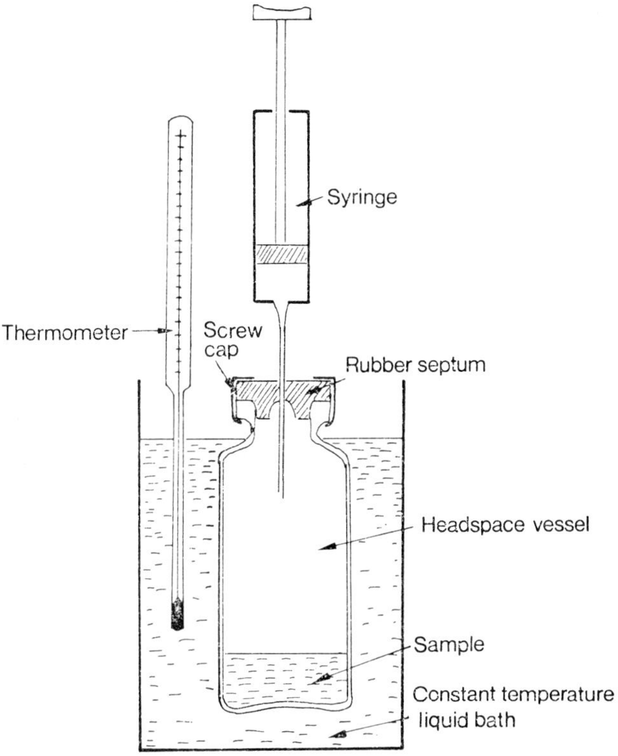

Fig. 11 Headspace analysis vessel.

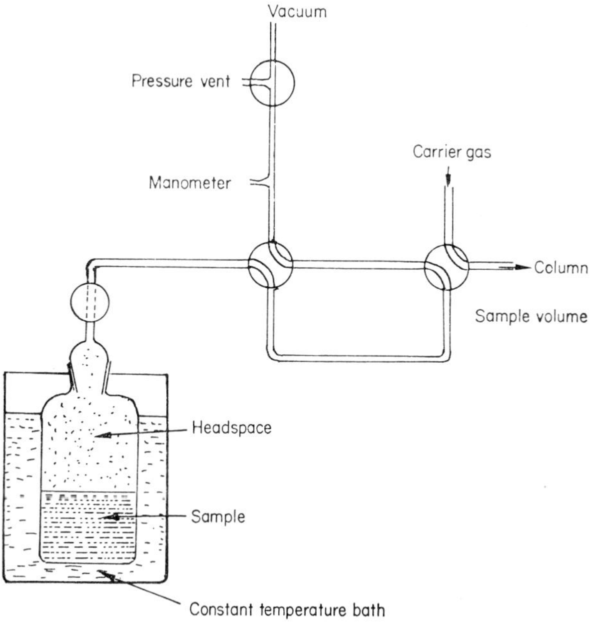

Fig. 12 Principle of sampling by means of reduced pressure (partial pressure sampling of vapours).

(3) To facilitate effective and rapid temperature equilibration, the volumes of the headspace analysis vessel and the headspace itself should not be too large. A glass vessel of about 10 cm^3 capacity is generally used to contain a sample of 2–3 cm^3 or g. Care must be taken to ensure that the glass is sufficiently strong to withstand high pressures if necessary.

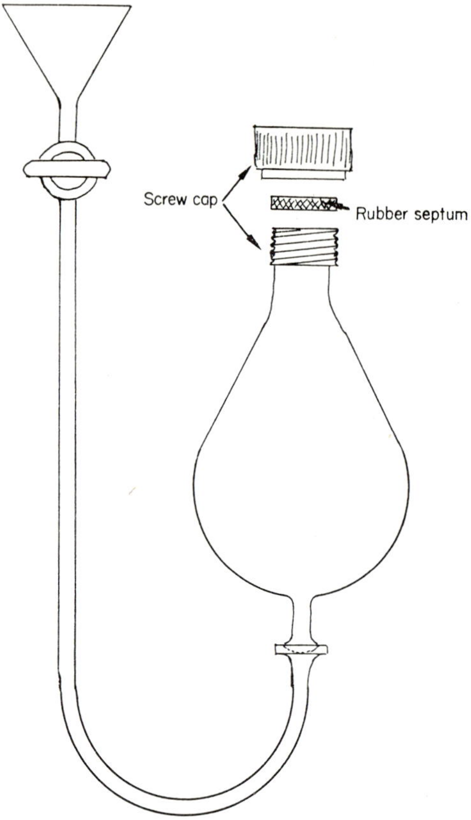

Fig. 13 Headspace bulb (from W. Piorr, *GIT*, **12**, 175 (1968): Fig. 1).

A device specially developed for gas chromatographic headspace analysis is the 'headspace bulb'[22] (Fig. 13), which differs from the usual apparatus in enabling headspace analysis to be performed at constant pressure.

Figure 13 shows that the device consists of a pear-shaped glass bulb with a means of removing samples at its taper end. The lower end is connected through a ball-and-socket joint to a pressure equilibration tube fitted with a glass stopcock and funnel. The preparation of the specimen before removing the sample with a syringe is as follows. The pressure equilibration tube is filled with water up to the ball joint connection. The headspace bulb is then completely filled with the liquid specimen, the top of the bulb is sealed by securing the rubber septum and the stopcock in the pressure equilibration tube is opened. About

2–5 cm^3 of an inert gas, e.g. nitrogen or helium is then blown into the headspace. After the vapour pressure equilibrium has been allowed to establish, a syringe is used to remove about 0.5–1 cm^3 as a sample for gas chromatographic analysis. It is advisable to use a syringe which can be heated, and, if possible, it should be at the same temperature, or even a higher temperature, than the vapour phase in the headspace.

According to Piorr, this experimental technique offers a reliable basis for quantitative analysis and has the following advantages: (a) a constant gas pressure, which is independent of the experimental temperature, can be maintained in the headspace which results in (b) a wide choice of working temperatures; (c) a headspace bulb of appropriate volume can be used for each analytical problem; and (d) a constant headspace volume can be obtained by using a gas-tight syringe.

1.53. Manipulation of the Sample

While being transported to the analytical laboratory the vessel containing the specimen must be completely full, so that there is no gas phase present. The stopper must form an absolute seal, and corks or rubber bungs should not be used. Ground glass stoppered flasks are the most suitable, provided they are completely filled up to the stopper.

During transfer of the specimen into the actual headspace analysis vessel, care must be taken to ensure that none of the readily volatile components are lost by evaporation. It is not satisfactory to simply pour the specimen, especially if only a small quantity is available. Even transfer by means of a syringe involves the risk of fractionation and condensation as a result of the compression and expansion by the plunger. The most satisfactory method is to use a siphon so that the specimen can be transferred into the headspace analysis vessel without a change in the pressure. The actual removal of the sample from the headspace for gas chromatographic analysis is the most important step in the quantitative determination. There are several ways of doing this, some of which differ significantly from the standard practice in gas analysis.[23] Either a direct or an indirect method of sampling may be used, depending on the concentrations of the components present in the headspace, although under certain circumstances these may be increased by one of the methods mentioned in Section 1.3. Indirect sampling is always preceded by an enrichment step, and must be employed when very low concentrations are present in the headspace vapours.

1.54. Headspace Sampling

1.541. Combined sampling and enrichment by condensation or chromatographic sorption[24] (indirect sampling)

In this method a known amount of vapour is removed from the headspace by means of a syringe, a capillary tube cooled to a low temperature or a short gas–solid or gas–liquid chromatographic column being interposed between the

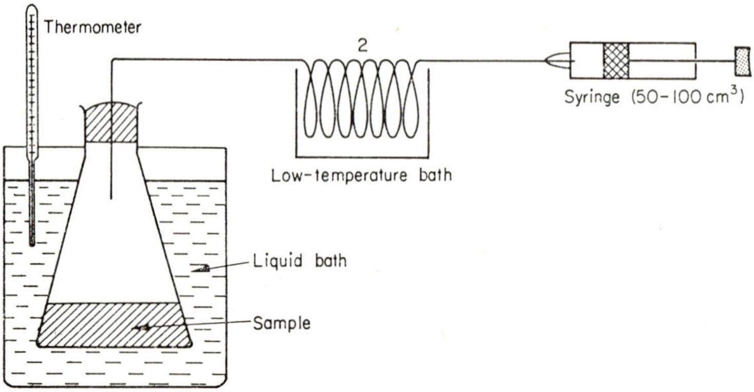

Fig. 14 Sampling from the headspace using an enrichment step.

specimen container and the syringe.[25] Figure 14 shows an arrangement of this kind in which the sample is withdrawn from the vapour phase in the specimen container through a cooled capillary tube by means of the syringe, which is also used to measure the volume removed. The method of transferring the enriched components on to the gas chromatographic column is evident from Fig. 15.

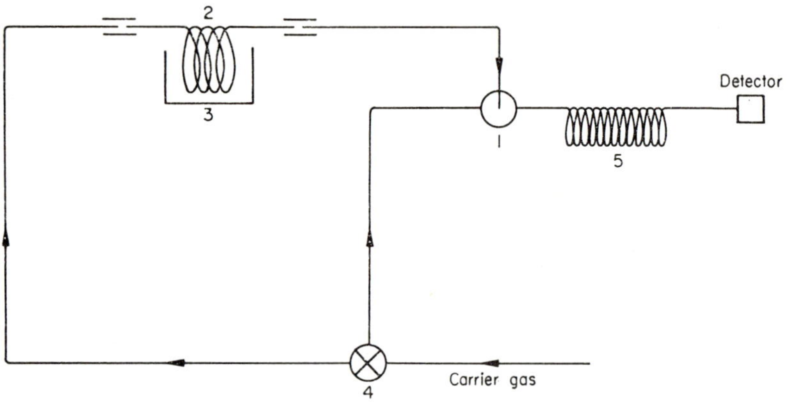

Fig. 15 Transfer to the chromatographic column of samples taken from the headspace (from E. Sprecher *et al.*, *Z. Naturforsch.*, **18b**, 495 (1963): Fig. 2).

The capillary tube 2 (see Fig. 15) is connected to the injection block 1 of the gas chromatograph and to a multiway valve 4 in such a way that the contents of the capillary can be swept into the column 5 with the carrier gas passing through 4. During transfer the capillary tube can be heated by immersing it in the liquid bath 3.

This method of transferring the sample to the column can also be performed with a gas sampling valve; however, it should be of a type that can be heated. In this case the procedure is obvious from Fig. 16.[26]

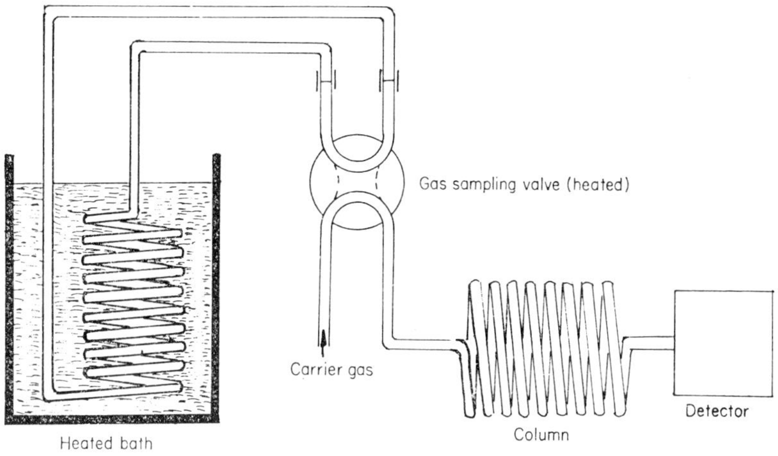

Fig. 16 Method of using a gas sampling valve to transfer a headspace sample concentrated in a capillary tube (from F. Drawert *et al.*, *Chromatographia,* **2,** 77 (1969): Fig. 1).

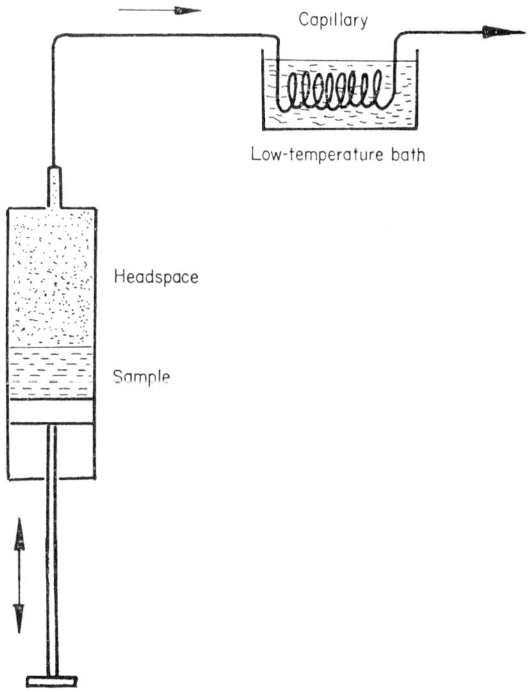

Fig. 17 Concentration of the headspace components with a syringe.

Another elegant approach to this problem consists of employing the syringe itself as the headspace analysis vessel.[27] This is done by introducing the sample into a 100 cm^3 all-glass syringe with its tip pointing upwards. After being shaken to establish the equilibrium, the cooled capillary tube is connected to the upper end and the gas phase transferred into the capillary tube by moving the syringe plunger upwards as indicated in Fig. 17. The concentrated components are subsequently transferred to the column by one of the methods illustrated in Figs. 15 and 16.

The actual sampling procedure can also be carried out in such a way that the loaded collection tube is connected directly to the chromatographic column by means of a screw coupling, and so becomes an integral part of the gas chromatograph.[28] However, this technique is cumbersome and time-consuming, since before each sample is collected the column must be taken apart and reassembled. This can be avoided by carrying out the concentration step in the initial section

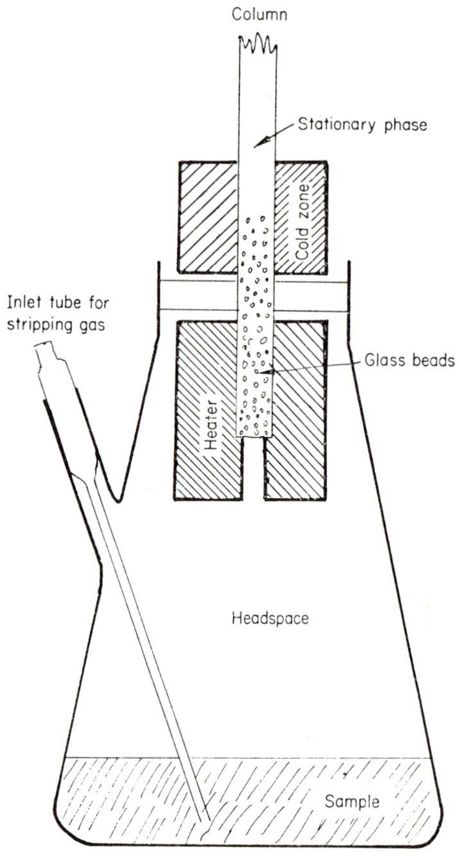

Fig. 18 Sampling from the headspace (from H. Binder, *J. Chromatog.*, **25**, 189 (1966): Fig. 1).

of the chromatographic column itself.[29] This results in the sample volume being compressed so that the substances being analysed enter the column in a narrow zone, and they can be further concentrated if necessary by cooling. Since the loaded pre-column is incorporated into the gas chromatograph, the analysis can be performed in a single operation under temperature programmed conditions. Binder[30] developed the apparatus shown in Fig. 18 for this technique. The initial 3–5 cm length of the column is filled with uncoated glass beads (0.2 mm diameter). This gives a short starting zone at the beginning of the actual chromatographic column. The transfer of the components in the vapour phase occurs as a result of the inert gas flow at the column inlet rather than as a consequence of the lower pressure at the column outlet.

It is recommended that for this application the sintered metal disc which is often fitted in commercially produced columns should be replaced by glass wool in order to provide the least possible resistance to the passage of material from the headspace to the stationary phase in the column.

Another way of performing this combined sampling and concentration step is the circulation method[31] shown in Fig. 19.

The advantage of this method lies in the thorough removal of the volatile components from the specimen being analysed, and this is particularly important, for example, when the components have to be separately identified.

The concentration techniques described above sometimes involve errors and difficulties, particularly in quantitative analysis, as a result of the occurrence of fractionation processes. If the possibility of forming azeotropic mixtures is disregarded, then in these concentration processes the readily volatile components will, of course, always be collected first and be followed by the difficultly volatile components, the relative volatilities being determined by the solvent.

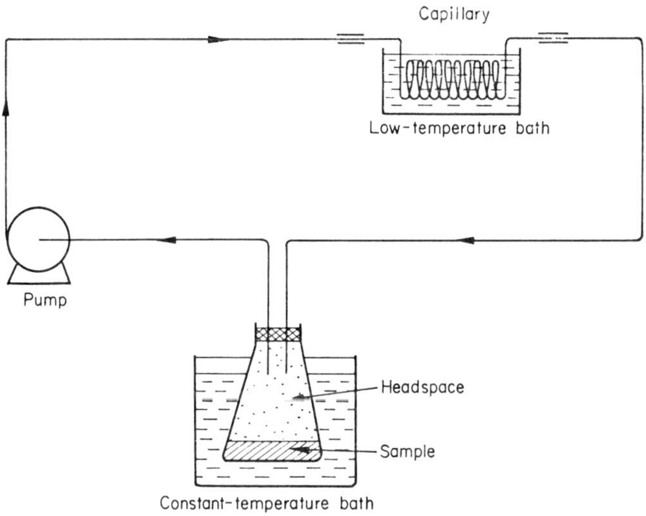

Fig. 19 Circulation method of concentrating headspace components.

Quantitative analyses based on these methods of sampling are very difficult, and in many cases they are even impossible. This is illustrated by the comparison made by Nawar and Fagerson[31] of the circulation method of concentrating components with direct sampling from the headspace for the analysis of a homologous series of aliphatic ketones dissolved in water or in an organic medium (corn oil) (Table 5).

TABLE 5

Comparison of peak areas obtained by direct sampling from the headspace and by the enrichment technique (sample: 200 p.p.m. acetone, 2-butanone, 2-pentanone, 2-hexanone, 2-heptanone in water and corn oil)

Compound	*Peak area by direct sampling*		*Peak area after concentration step*		*Enrichment factor*	
	aqueous phase	*organic phase*	*aqueous phase*	*organic phase*	*aqueous phase*	*organic phase*
Acetone	10	47.0	268	2547	26.8	54.2
2-Butanone	14	11.0	298	707	21.3	64.3
2-Pentanone	22	5.7	321	338	14.6	59.3
2-Hexanone	29	2.7	119	78	4.1	29.0
2-Heptanone	24	0.7	26	11	1.1	15.0

The accumulation of relatively large amounts of water often causes difficulty with the concentration step. This can be avoided to some extent by selectively removing the water during the concentration procedure by means of cold traps or drying agents, or by using hydrophobic polymers in the pre-column in which the enrichment takes place.[32,33] The following methods may be considered for the actual concentration process in capillary tubes.

(1) Simple condensation in cooled capillary tubes. Since this method is not very selective and does not achieve very high degrees of enrichment, it is not employed very frequently. Moreover, a delayed vaporization of the condensate when the sample is introduced into the chromatograph can cause marked tailing of the peaks.

(2) Adsorption on silica gel, activated charcoal and similar materials. This is primarily applicable to low-boiling non-polar materials, since chemical changes can take place with polar substances and invalidate the results of qualitative and quantitative analyses.

(3) The phase equilibration method.[1] The principle of this method of concentration, which is used when polar and high-boiling materials are present, depends on using a small amount of a gas–liquid chromatographic column packing at a definite temperature to trap the sample from the vapour phase. Equilibrium is established between the vapour phase and the liquid phase in a single frontal chromatographic process. The enrichment in the liquid phase occurs in direct proportion to the value of the corresponding partition coefficient.

On this basis Vitenberg, Ioffe and Borisov[34] have recently developed an excellent method for the selective concentration and determination of atmospheric pollutants. This involves saturating the liquid stationary phase with the pollutants concerned until equilibrium has been established. Subsequently these constituents are determined in the headspace above their solution in the liquid stationary phase. Activated charcoal can be used as a suitable column packing for the headspace analysis.[121] It is generally true of these enrichment procedures that they involve too much effort, that they are time-consuming and that they are prone to too many errors as far as quantitative measurements are concerned. They therefore find only limited application in the field of routine analysis. Sample injection from the headspace directly on to the gas chromatographic column is therefore to be preferred if possible.[1]

1.542. Direct sampling from the headspace

Direct sampling from the headspace can be performed with normal syringes, by partial pressure sampling and by other special sampling techniques.

The advantages and disadvantages of sample removal and injection with syringes are obvious, but it must be remembered that the techniques give rise to a very wide range of potential errors. It is essential that the syringe be absolutely grease-free and at a higher temperature than the headspace sample.[35] After the

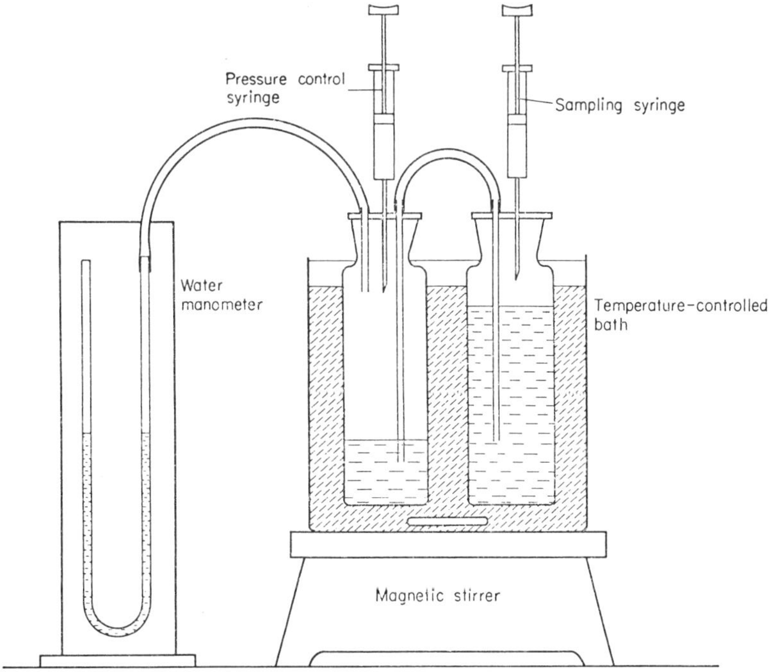

Fig. 20 Sampling with pressure equalization (from P. Ronkainen, *Kem. Teollismus*, **26**, 215 (1969)).

sample has been injected, the syringe must be carefully cleaned by pumping under vacuum at a higher temperature. The back-pressure in the injection block should be as low as possible, to permit rapid injection. By taking all of these precautions it is possible to perform quite rapid analyses by this method if the concentrations in the gas phase are sufficiently high. However, the quantitative results are usually not entirely satisfactory. This type of sampling should therefore be employed only in exceptional cases.

A special technique for injecting with a syringe and which largely avoids errors due to pressure differences during sampling, has been developed by Ronkainen.[36] Figure 20 illustrates the principle of the method, in which the pressure control syringe and the manometer can be used to adjust the pressure so as to keep it constant during removal of the sample. This minimizes the possibility of fractionation and condensation processes occurring.

Another method of sampling developed specifically for gas chromatographic headspace analysis involves transferring the sample into a constant volume by using a reduced pressure. The apparatus shown in Fig. 21 was developed to use this method for the analysis of the gases contained in bottled beer.[37]

In contrast to normal gas analysis practice, the sampling procedure illustrated in Fig. 21 requires that the gas sampling valve and the associated tubing be heated when vapours are being analysed. In the form illustrated it is therefore

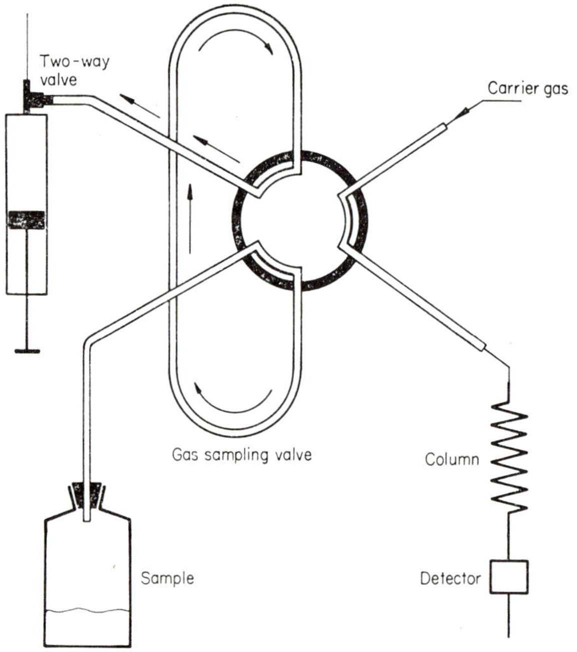

Fig. 21 Sampling from a headspace by means of a reduced pressure system (from H. Zenz *et al.*, *Mitt. Versuchssta. Gärungsgewerbe*, **22**, 175 (1968)).

only applicable to gases and very low boiling point compounds. The most reliable method of performing accurate quantitative gas chromatographic headspace analysis is that employed in the GC-Automat Multifract® chromatograph model F42 (Perkin-Elmer GmbH, Bodenseewerk, West Germany).[38] In this instrument the sample is introduced directly from the headspace analysis vessel into the chromatographic column by means of an excess pressure. The instrument can be equipped with flame ionization, thermal conductivity, electron capture and other detectors, and functions completely automatically, so that operator errors, especially those involved in sampling, do not arise. The procedure for carrying out an analysis is as follows.

The specimen is introduced into a serum bottle having a volume of about 30 cm^3. This is then hermetically sealed by means of a rubber septum and aluminium foil, which is pressed on to the serum with a special pair of pliers. A thermostat capable of holding 30 of these samples ensures that they all reach the same equilibrium temperature, and they are then brought in succession to the automatic injector by means of a turntable. At the beginning of an analysis the turntable is raised and a hypodermic needle penetrates the rubber septum on the serum bottle. At the same time the carrier gas pressure, which is controlled by a regulator, rises until it equals that at the column inlet. The time required for this is preset and can not be varied, and when it is completed the carrier gas valve closes automatically. The excess pressure in the serum bottle is slowly released via the column, the volatile constituents in the sample being simultaneously carried into the column. At the end of this sampling sequence the valve on the automatic injector closes and the carrier gas supply to the column is resumed.

It is arranged that when the sampling period is completed the turntable lowers and additional carrier gas is blown through the automatic injector via another

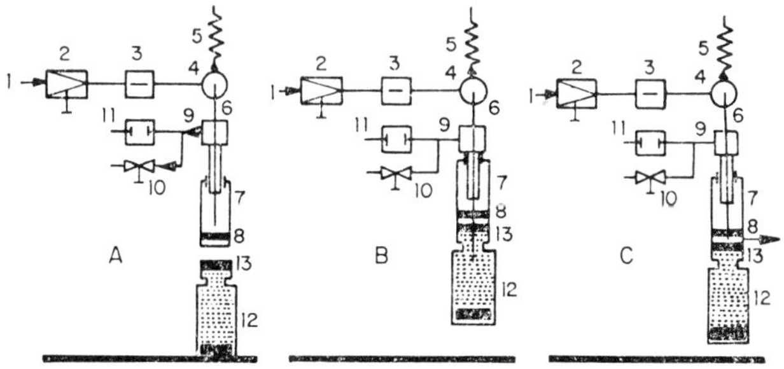

Fig. 22 Schematic diagram of the sampling system of the Multifract® instrument model F42, Perkin-Elmer, Bodenseewerk: 1, carrier gas inlet; 2, pressure regulator; 3, magnetic sampling valve; 4, inlet; 5, column; 6, heated sampling capillary; 7, movable cut-off cylinder; 8, silicone rubber septum; 9, flushing line; 10, variable pressure regulator; 11, magnetic relief valve; 12, specimen bottle; 13, rubber septum cap: A, sampling system in rest position; B, sampling system and specimen bottle in the 'sampling' position; C, sampling system and specimen bottle in the 'flushing' position.

Fig. 23 Thermostat containing specimen bottles, and the sampling unit of the F42 Multifract instrument (Perkin-Elmer, Bodenseewerk).

Fig. 24 General view of the Multifract F42 chromatograph (Perkin-Elmer, Bodenseewerk).

valve. This removes the less volatile components which would otherwise be carried over into the next analysis. The duration of this flushing can be extended or shortened as required. When this sequence of operations has finished, the analysis of the next sample commences. The instrument automatically stops after the 30 samples. Figure 22 is a schematic diagram of this sampling system, and also illustrates the mode of operation. Figure 23 shows the thermostat containing the sample bottles and the sampling unit, and Fig. 24 is a general view of the F42 instrument.

Pauschmann[39] has developed a similar method based on the principle of excess pressure sampling, which can be employed with all gas chromatographs. A Teflon tube is connected to a T-piece inserted into the normal carrier gas connections of a gas chromatograph, and leads to a metal stopcock having a narrow-bore ground taper for attaching a needle. This arrangement enables a higher carrier gas pressure to be maintained in this section than is used in the actual analysis. Instead of the usual injection septum the injection port is fitted with a capillary tube which also leads to a metal stopcock. The principle of this system is shown in Fig. 25.

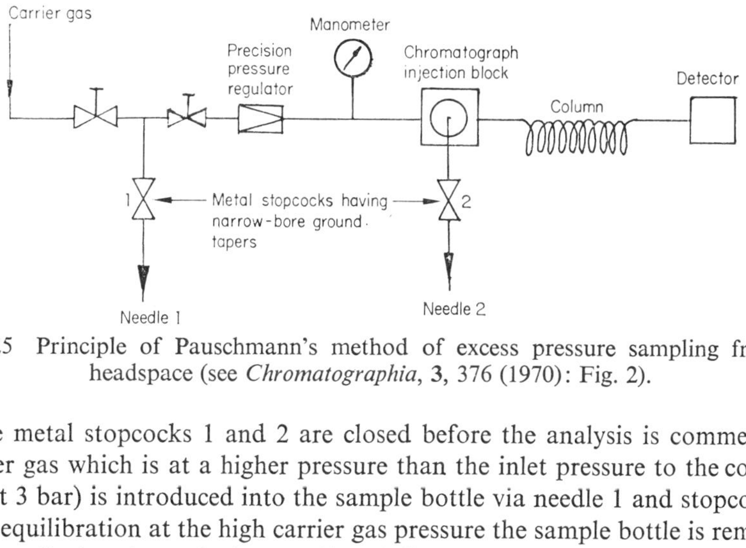

Fig. 25 Principle of Pauschmann's method of excess pressure sampling from a headspace (see *Chromatographia*, 3, 376 (1970): Fig. 2).

The metal stopcocks 1 and 2 are closed before the analysis is commenced. Carrier gas which is at a higher pressure than the inlet pressure to the column (about 3 bar) is introduced into the sample bottle via needle 1 and stopcock 1. After equilibration at the high carrier gas pressure the sample bottle is removed from needle 1 and attached to needle 2. When the metal stopcock 2 is opened, the vapour phase sample is forced into the column by the excess pressure, which must, of course, be higher than that of the carrier gas.

Ordinary medicine bottles of about 10 cm^3 capacity can be used as sample containers. The injection block is so constructed that the normal silicone rubber septa can still be used for the ordinary gas chromatographic analysis of liquid samples.

Another sampling system based on the excess pressure principle, and which can be employed on gas chromatographs fitted with dual injection ports, has been described by Göke.[40] The mode of operation can be seen from Fig. 26. Only one column is installed in the dual-column instrument, a blanking plug being fitted into the second column connection. The pressure in the headspace analysis vessel (see Fig. 26) builds up via the metal stopcock A when C is closed.

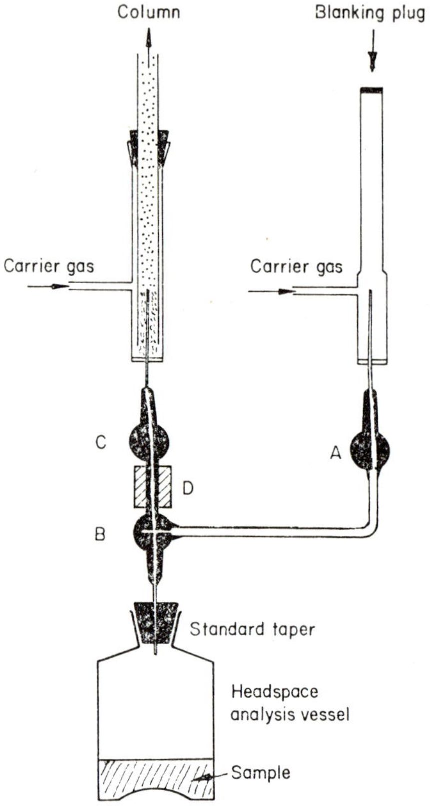

Fig. 26 Sampling method used by Göke for gas chromatographic headspace analysis (from G. Göke, *Chromatographia*, **5**, 622 (1972)).

Care must, of course, be taken to ensure that this pressure is higher than that at the column inlet. Injection of the sample is then achieved by closing A, rotating B and opening C. The advantages of this system compared with that in Fig. 25 are that the sample vessel does not need to be changed, and the heated section D ensures that the sampling unit from C to the point at which the sample is attached is efficiently purged at the end of the analysis.

1.55. The Sources of Error in Gas Chromatographic Headspace Analysis

As indicated in the preceding section, the most important requirement for quantitative headspace analysis is the stable performance of the instrument employed. It is also essential to know the time required for the equilibrium vapour pressures to be established. Highly viscous samples generally require a longer period for this than low-viscosity samples. The large differences in the rates of diffusion of the individual substances into the vapour phase can be the principal cause of errors if the time allowed for equilibration is too short and if the headspace volume is too large. Binder avoided this problem by mounting a rotating impeller in the headspace, its pivot being on a ground-in lid which also served as a stopper.[23] An alternative approach is to pass a slow stream of gas through the sample and use a syringe or gas-sampling valve to remove small fractions of it; however, the size and shape of the gas bubbles influences the equilibrium concentrations which are established.[41] Sampling from containers sealed with rubber septum caps can involve uncertainties which can significantly affect the reproducibility and accuracy of the analysis.[42] For example, the type of rubber used in the cap is very important, since considerable amounts of the trace components being analysed can be absorbed by rubber septa. Figure 27 shows the increasing amounts of various organic vapours absorbed in rubber septa as a function of time. Maier has demonstrated that this absorption cannot be entirely prevented by heating the septum or covering it with aluminium foil or Teflon film, and experiments with silvered septum caps also gave unsatisfactory results.[43] Davis considers that this absorption in the material used to seal the headspace is a principal source of error, and it depends on the molecular weight and molecular structure of the compounds being analysed.[44] Thus, when

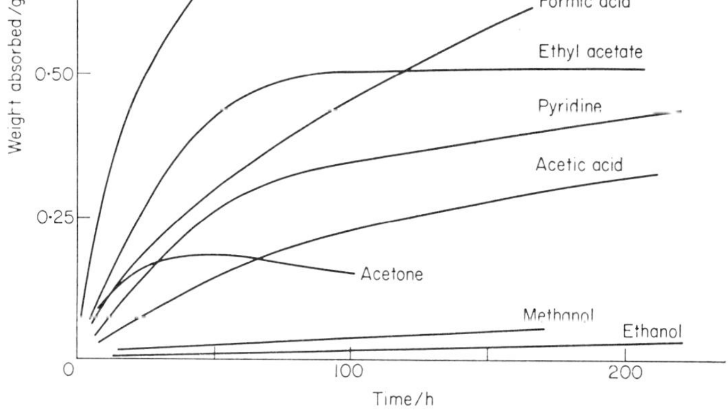

Fig. 27 Absorption of organic vapours in rubber septum caps at 23°C (from H. G. Maier, *J. Chromatog.*, **50**, 329 (1970)).

a rubber stopper is used to seal the headspace, the concentrations of the vapours in it can decrease within 30 min by 2% for ethylene, 7.6% for hexane and 21.9% for heptane. The loss is even greater for aldehydes, being 4.6% for propionaldehyde, 26.3% for valeraldehyde and 64.5% for heptanal. Davis therefore recommended that glass stopcocks be used, since they eliminate this effect. However, since it is not usually possible to avoid using a rubber septum, it is advisable to determine a correction factor for the septum and take this into account when calculating the analytical result.

1.6. PRACTICAL APPLICATIONS OF HEADSPACE ANALYSIS

1.61. General

Gas chromatographic headspace analysis is concerned predominantly with the determination of traces of volatile substances in samples which are difficult or impossible to handle by conventional gas chromatographic methods. These include liquid and solid materials and mixtures, on which it is difficult to perform quantitative analyses or to interpret their results. The materials which have to be considered can be classified as follows:

Group I. Homogeneous liquid samples. For example: sewage, drinking water, beverages, spirits, vegetable oils, mineral oils, etc.
Group II. Inhomogeneous samples consisting of a liquid and a solid. For example: blood, milk, polymer dispersions, etc.
Group III. Solid materials which give homogeneous solutions. For example: soluble polymers, inorganic salts, etc.
Group IV. Insoluble solid samples. For example: high molecular weight olefins, foodstuffs, fruits, tobacco, spices, etc.

Whereas for Groups I, II and to some extent Group III it is possible to obtain accurate quantitative results for the volatile constituents being determined, the materials in Group IV can still only be characterized by means of comparative chromatograms, i.e. 'fingerprints'.

1.62. Group I—Homogeneous Liquid Samples

1.621. Volatile substances in water

An important application of gas chromatographic trace analysis is the determination of volatile trace constituents in drinking-water and sewage. The problems which arise in the classical gas chromatographic analysis of these materials, e.g. overloading of the column and the detector by the main component, water, and the occurrence of memory effects, are well known to analysts. However, they do not represent a serious problem in gas chromatographic headspace analysis, since, depending on the temperature of the aqueous sample, at most about 2–4% by volume of water vapour reaches the column and the detector.

In a fundamental study of the gas chromatographic headspace analysis of aqueous solutions for traces of alcohols, ketones, aldehydes, esters and sulfides in the range 0.01–10 p.p.m. in water Özeris and Basette[45] found a linear relationship between peak height and concentration as shown in Fig. 28.

The slopes of the straight lines in Fig. 28 depend on the type of compound and define their limits of detection. While most of these compounds can be analysed down to 0.01 p.p.m., alcohols cannot be determined below 0.1 p.p.m., and

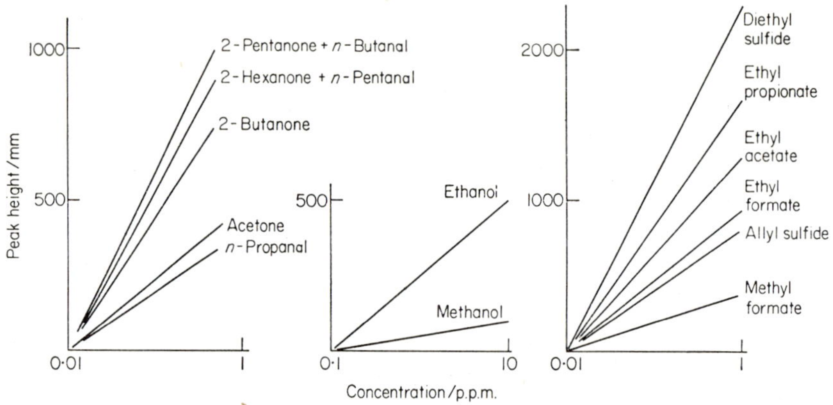

Fig. 28 Peak heights as a function of concentration for various organic compounds (from S. Özeris and R. Basette, *Anal. Chem.*, **35**, 1091 (1963)).

methanol not even below 1 p.p.m. (For activity coefficients in water see Fig. 7.) The limits of detection can be lowered still further by adding inorganic salts, as already discussed in Section 1.3, the addition of sodium sulfate during the analysis of carbonyl compounds being an example.[15] Kepner, Maarse and Strating obtained up to a sevenfold increase in sensitivity by adding ammonium sulfate or sodium chloride until the dilute aqueous sample was saturated. The equilibrium vapour pressure was established by agitating the solution at constant temperature.[46] A linear dependence of peak height on the amounts of various alcohols and esters was also obtained by these authors, who used an internal standard in the analyses. Jentzsch, Krüger and Lebrecht showed that a significant increase in sensitivity towards *tert*-butanol in aqueous solution can be obtained by adding calcium carbonate.[47] The limit of detection was reported to be 0.05 p.p.m.

Gottauf[27] has shown that compounds can be detected at the level of $10^{-9}\%$ by combining the salting-out technique with an enrichment step. Table 6 gives some limits of detection in water for this method applied to odours derived from foodstuffs.

These low limits of detection make it possible to determine traces of odours which migrate from one foodstuff to another during normal storage in household

TABLE 6

The effect of the type of salt and its concentration on the limits of detection in water for odours from foodstuffs, as reported by Gottauf

Compound	*Salt solution (containing* 100 cm^3 H_2O)	*Limit of detection/*%
2-Pentanone	45 g Na_2SO_4, 0.71 g KH_2PO_4	6×10^{-8}
Hexanal	1.42 g Na_2HPO_4, $2H_2O$	5×10^{-8}
Isoamyl alcohol	1.42 g Na_2HPO_4, $2H_2O$	8×10^{-7}
Limonene	1.42 g Na_2HPO_4, $2H_2O$	7×10^{-9}
Hexyl acetate	1.42 g Na_2HPO_4, $2H_2O$	2×10^{-8}
Dipropyl disulfide	1.42 g Na_2HPO_4, $2H_2O$	2×10^{-8}
Trimethylamine	45 g Na_2SO_4, 0.4 g NaOH	5×10^{-8}
Trimethylamine	100 g NaOH	6×10^{-9}

refrigerators. Figure 29 shows the headspace chromatograms for a sample of water which had been exposed for several minutes to the odours of different foodstuffs in a household refrigerator.

The determination of trimethylamine was carried out in an acidified Na_2SO_4 solution (chromatogram A2), and that of the two aldehydes in a hydroxylammonium sulfate solution (chromatogram B2). Other examples of the analysis of water are the determination of chlorodane (1,2,4,5,6,7,8,8-octachloro-2,3,3a,4,7,7a-hexahydro-4,7-methanoindene)[48] and of dichloromethane, cyclohexane and chloroform down to the level of about 0.1 p.p.m.[49] Traces of organic solvents which can get into drinking-water from the protective coating on potable water tanks may also be determined with very high sensitivity, and Miethke has reported the limits of detection given in Table 7.[42]

1.622. Trace components in alcoholic beverages

The effect on the vapour pressure equilibrium caused by a decrease in the solubility in aqueous solutions, such as that due to salting out, can, of course,

TABLE 7

Limits of detection for solvents in drinking-water

(from H. Miethke, *Deutsche Lebensmittelrundschau*, **65**, 379 (1969); Table 1)

Solvent	μl/l
Methanol	6.5
Ethanol	4.5
Acetone	0.3
Benzene	0.5
Toluene	1.0
Trichloroethylene	2.5
Light petroleum hydrocarbons	20

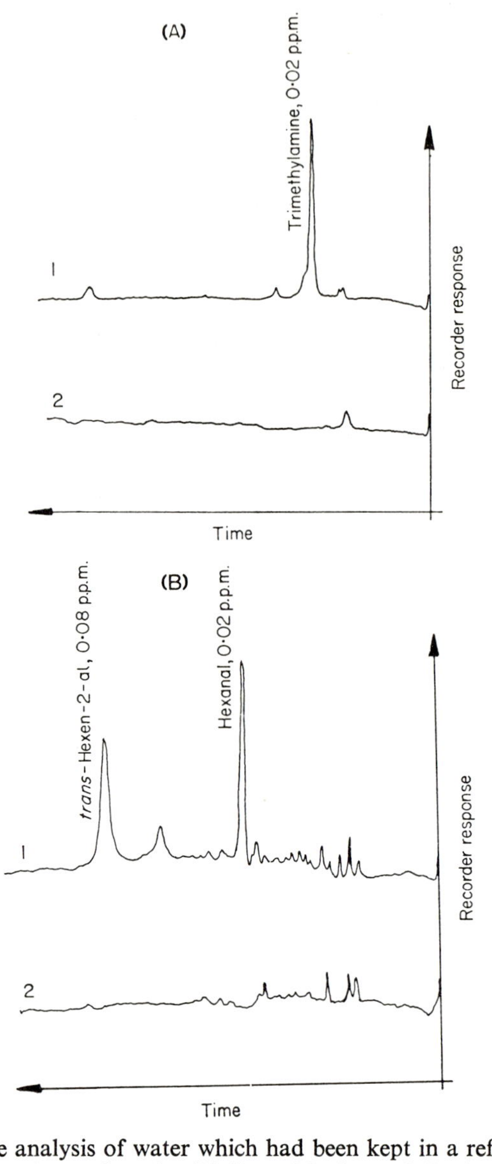

Fig. 29 Headspace analysis of water which had been kept in a refrigerator together with cod fillets (A) and grated apples (B) (from M. Gottauf, *Z. Anal. Chem.*, **218**, 175 (1966): Fig. 3).

also occur as a result of other compounds being present in the sample (see Section 1.3), and attention should be paid to this point in the analysis of alcoholic beverages. Large variations occur in the concentrations of these materials from one sample to another, and can lead to problems in the quantitative interpretation of results. Thus, for example, high concentrations of ethanol influence the solubility, and consequently the vapour pressures, of other compounds which are present at lower concentrations. Figure 30 shows how ethanol interferes in this way with the determination of amyl alcohols and esters.

In the case of beverages having a high alcohol content it is therefore advisable to dilute them to about 20% with water. It is then possible to determine, for example, 2-methyl propanol at concentrations ranging from 0.01 to 0.03 vol.% with a relative accuracy of ±10%, and methanol over the range 0.06–0.24 vol.% with an accuracy of 8%.[42] In the headspace analysis of beer and lemonade the large amounts of carbon dioxide present have to be taken into account, since

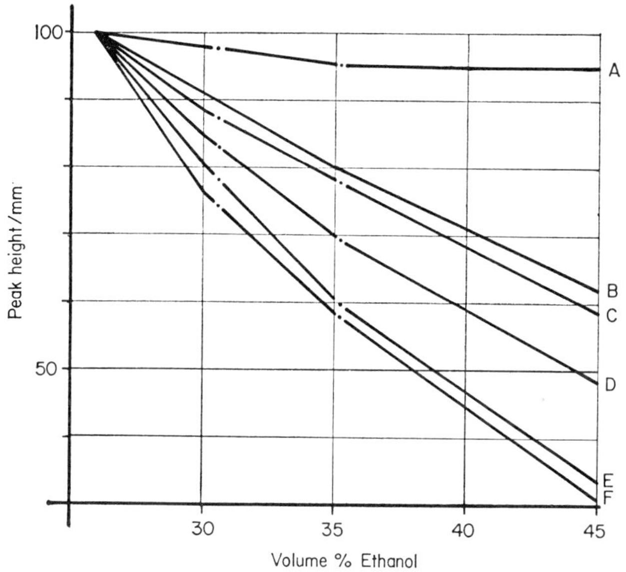

Fig. 30 Peak heights for some alcohols and esters as a function of the concentration of ethanol: A, methanol; B, ethyl acetate; C, *n*-propanol; D, *n*-butanol; E, isoamyl acetate; F, amyl alcohol from fermentation (from H. Miethke, *Deut. Lebensmittel-rundschau*, **65**, 379 (1969): Fig. 13).

they can produce the same effect as ethanol. Similar experiments have shown that the peak heights observed for isoamyl acetate, isobutanol and isoamyl alcohol in aqueous solution are smaller than in beer. On the other hand, some compounds are much less affected than those mentioned above. Thus, ethyl caproate gives the same peak height in either water or rum. This illustrates how difficult and time-consuming quantitative headspace analysis can be, since it means that a special calibration must be performed for each product being analysed. Thus, calibrations have to be performed in non-carbonated samples of beer for the quantitative determination of the fermentation by-products. The reproducibility obtained in this way for the determination of acetaldehyde, ethyl acetate, *n*-propanol, isobutanol, isoamyl acetate and the amyl alcohols is shown in Table 8.[50]

The standard deviations of the mean values of the results in Table 8 are: for acetaldehyde (a), ±2.4%; methyl acetate (b), ±0.8%; *n*-propanol (c), ±3.1%; isobutanol (d), ±1.1%; isoamyl acetate (e), ±1.2% and the amyl alcohols (f), ±0.8%.

Figure 31 shows the chromatogram for a headspace analysis of a sample of beer.

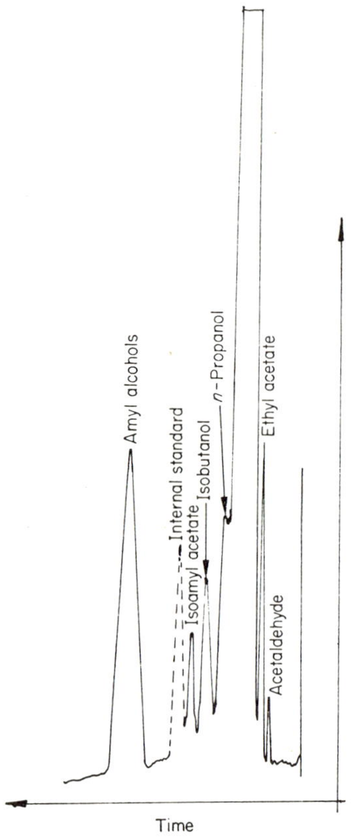

Fig. 31 Headspace analysis chromatogram showing the determination of fermentation by-products in a sample of beer (Carbowax® column in F40 instrument) (from B. Mändl *et al.*, *Brauwissenschaft*, **22**, 477 (1969): Fig. 3).

Headspace analysis using an electron capture detector for the determination of ketones such as diacetyl and pentane-2,3-dione in beer gives results in agreement with those obtained by the spectrophotometric method. This is demonstrated by the comparison of the determination of diacetyl in different types of beer given in Table 9.

In a similar study Jahnsen and Horn[51] established the presence of 26 constituents in beer, and termed the chromatogram an 'aromagram'. The following

TABLE 8

Reproducibility in the determination of fermentation by-products in beer
(from B. Mändl *et al.*, *Brauwissenschaft*, **22**, 477 (1969); Table 1)

Analysis	mg/l *a*	mg/l *b*	mg/l *c*	mg/l *d*	mg/l *e*	mg/l *f*
1	10.0	24.8	10.5	13.2	1.88	65.0
2	10.2	24.8	11.3	13.3	1.86	64.6
3	10.7	26.0	12.0	14.3	2.00	67.5
4	11.1	25.6	11.3	14.2	1.94	67.0
5	11.8	25.0	11.1	14.0	1.90	65.2
6	11.5	26.4	12.2	14.4	2.02	68.7
7	11.2	25.6	12.0	13.6	2.04	66.0
8	12.0	26.1	10.2	14.1	2.06	67.5
9	12.2	25.9	10.5	13.7	2.02	66.5
10	12.7	26.6	10.7	14.3	2.04	68.0
mean value	11.3	25.7	11.2	13.9	1.98	66.0

compounds were identified on the basis of retention data: dimethyl sulfide, acetaldehyde, formaldehyde, ethyl formate, acetone, ethyl acetate, methanol, ethanol, isopropanol, ethyl propionate, diacetyl, *s*-butanol, isobutyl acetate, *n*-propanol, ethyl butyrate, pentan-3-ol, isobutanol, 2-methyl butan-2-ol, isobutyl isobutyrate, isoamyl acetate, *n*-butanol, myrcene, isoamyl alcohol, ethyl caproate, *n*-pentanol, hexyl acetate, acetoin, ethyl lactate and ethyl caprylate.

Fingerprint chromatograms of beer which has been exposed to light have been used to investigate the influence of radiation from the sun and from fluorescent lights on its 'sunstruck flavour', the colour of the bottle being taken into account, since this is a contributing factor.[52] The stabilization of samples of beer containing yeast by pasteurizing them,[53] and also the changes in the composition of

TABLE 9

Comparison between headspace analysis and the spectrophotometric method for the determination of diacetyl in different types of beer
(from B. Mändl *et al.*, *Brauwissenschaft*, **22**, 477 (1969); Table 4)

Type of beer	*GC-headspace analysis* (mg/l)	*Spectrophotometric analysis* (mg/l)
Hell—lager	0.068	0.06
Pilsener	0.044	0.04
Hell—Export	0.072	0.07
Dunkel—Export	0.094	0.08
Weizen	0.095	0.05
Hell—Stark	0.046	0.07
Dunkel—Stark	0.044	0.04

substances producing aromas[54] as a result of heating, may be rapidly characterized by headspace analysis of the aroma constituents before and after these treatments.

The oxidation of aliphatic aminoacids in beer wort by ninhydrin to the corresponding aldehydes has also been studied by the same method.[55]

Another example which may be mentioned of gas chromatographic headspace analysis of the volatile constituents in beer is the determination of mycrene, alcohols, esters and carbonyl compounds.[56]

Miethke has carried out qualitative analyses of various kinds of alcoholic beverage, such as cherry brandy, brandy, perry, cognac, egg flip, etc., to identify them, their amyl alcohol constituents and adulterants.[42] The determination of the aroma constituents of rum,[57] and of ethanol in lemon juice[58] should also be mentioned here.

1.623. The determination of petrol constituents in engine oil

Headspace analysis provides a rapid and elegant test method for the monitoring of engine oils for petrol constituents.[59] This is demonstrated by comparing (Fig. 32) an unused engine oil (chromatogram A) with a used oil (chromatogram B).

1.624. The selective determination of vinyl chloride in olive oil

Another important example of the gas chromatographic headspace analysis of homogeneous liquid samples is the selective detection and quantitative determination of vinyl chloride in olive oil which has been stored in PVC containers.[60] The detector employed was a mass spectrometer focused on a single mass peak.[61] In order to use this method it is essential to know the mass spectrum of the compound being determined. The technique can be employed on a routine basis when traces of known compounds have to be determined in a

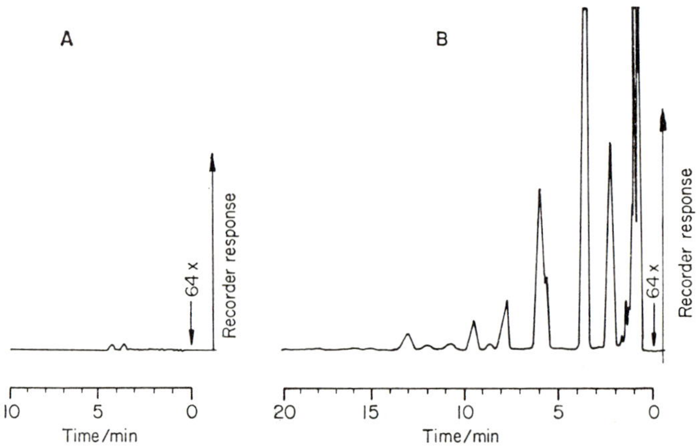

Fig. 32 Characterization of unused (A) and used (B) engine oil by headspace analysis.

Fig. 33 Mass spectrum of vinyl chloride (70 eV).

mixture of substances. It is particularly convenient if there is a very intense parent molecule peak which can be selected for the analysis. The mass spectrum reproduced in Fig. 33 shows that this is the case with vinyl chloride, the detection of which in foodstuffs is of the greatest importance. Figure 34 illustrates the selective analysis of a known mixture of 0.1 p.p.m. by weight of vinyl chloride in olive oil, the mass spectrometer being focused on mass 62 (molecular peak of vinyl chloride). The corresponding blank analysis of the empty headspace vessel is reproduced in Fig. 35. It is evident from Fig. 34 that the limit of detection of vinyl chloride by this method is <0.025 p.p.m. by weight. The experimental difficulties involved in this procedure should not be underestimated, since it involves continual checking of the precision of focusing on the parent molecule peak.

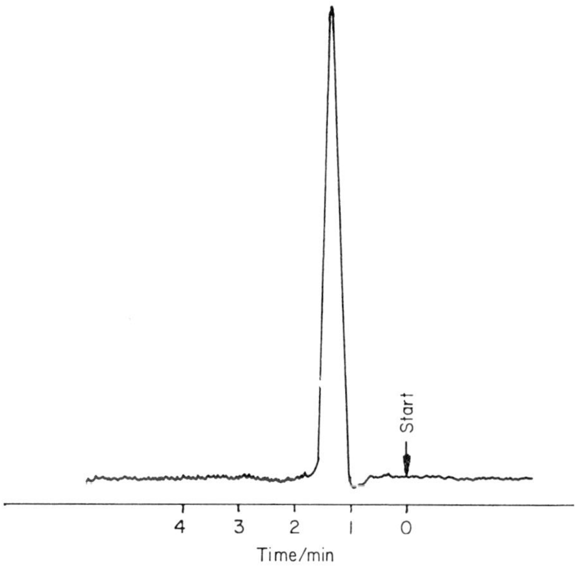

Fig. 34 The determination of 0.1 p.p.m. by weight of vinyl chloride in olive oil using a mass spectrometer as the detector.

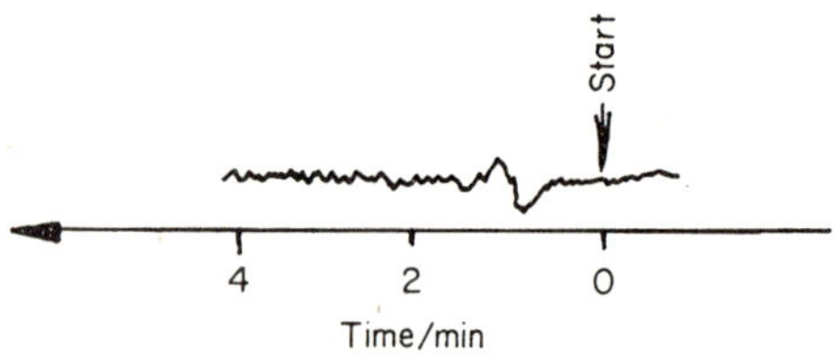

Fig. 35 Blank measurement on the empty headspace analysis vessel.

The experimental conditions for Figs. 33–35 were as follows.

Gas chromatograph–mass spectrometer assembly
GC: Perkin-Elmer model F20FE
MS: Perkin-Elmer model RMU6H
Gas sampling system fitted with 10 cm^3 sample loop; partial pressure sampling from the headspace analysis vessel
Column length: 3 m
Column diameter: 2 mm
Column packing: 10% polyphenyl ether + 2% Carbowax 20® on Chromosorb G®, 80/100 mesh
Column temperature: 25°C
Sample size from the headspace: 5 cm^3
Carrier gas: He
Carrier gas flow rate: 16 $cm^3\ min^{-1}$
Carrier gas flow rate to mass spectrometer (through flow restrictor): 1 $cm^3\ min^{-1}$
Recorder: Phillips PR2210, 50 mV
Focussed peak m/e: 62, at 20 eV (selected from the mass spectrum of vinyl chloride at 70 eV)
Retention time for vinyl chloride: 1.2 min
Sample preparation: 4 g olive oil containing 0.1 p.p.m. by weight of vinyl chloride held at 90°C in a water-bath under a He pressure of 2 kp cm^{-2}

1.63. Group II—Inhomogeneous Samples Consisting of a Liquid and a Solid Phase

While for materials in Group I a relatively simple two-phase liquid–gas system is present, in this group mass transfer processes can take place between the three phases—solid, liquid and gas. Attempts should therefore be made to dissolve samples of materials in this group in suitable solvents to give homogeneous solutions. This can be done, for example, with aqueous polymer dispersions. However, it is a much more difficult problem with natural products such as blood, serum, milk, etc., which are extremely difficult or impossible to convert into homogeneous solutions. For the quantitative analysis of these materials the medium itself must be available in a pure form, since otherwise it is necessary to depend on the peak enhancement method.

1.631. Residual monomers and other trace components in polymer dispersions

THE DETERMINATION OF BUTYL ACRYLATE, BUTYL PROPIONATE AND STYRENE IN AQUEOUS DISPERSIONS OF STYRENE–ACRYLATE COPOLYMERS[60]

Knowledge of the amounts of monomeric residues in aqueous dispersions of polymers is of importance with respect to their stability, and because they cause odours in the subsequent processing of the materials. The normal gas chromatographic analysis of the impurities in these dispersions is subject to many difficulties, since in addition to a solid, i.e. the polymer, there is also a large amount of water present. Although techniques such as isolation of the impurities by steam or azeotropic distillation, flushing the impurities being determined from a pre-column into the chromatographic column, or removal of the water by chemical reaction with acetic anhydride, dimethoxypropane, etc.,[1] exist, the direct sampling of aqueous dispersions by means of a syringe is still widely used. For this last method the sample must be diluted in the ratio 1:1 with water or an acetone–water mixture. In addition to the risk of demulsification occurring, and the difficulty of adding an internal standard, the technique has the following disadvantages.

(1) The solid polymer remains in the injection block of the gas chromatograph. Frequent replacement of the glass insert is essential if contamination is to be avoided, but this requires time.

(2) Sampling of a two-phase system—in particular, of polymers in water—with a syringe is an unsatisfactory method, since an internal standard cannot be employed. Quantitative analysis in this case therefore depends on the use of an external standard, and involves reproducible sampling and comparison with a standard mixture. In contrast with the internal standard method, this procedure should in principle avoid the more serious errors, especially in multiphase systems.

(3) The normal analytical technique used previously usually employed temperature programming, and this gives rise to poor baseline stability. Moreover, the relatively large amounts of water (about 50%) which elute with the peaks cause irreproducible memory effects which are clearly detrimental to trace analysis.

The gas chromatographic headspace analysis of homogeneous solutions of polymer dispersions in DMF avoids these problems. However, they are replaced by some others. For example, the solvent must be gas chromatographically pure, i.e. it must not contain any impurities which overlap the trace component peaks on the chromatogram. This also applies to the solvent peak. The search for a suitable solvent which fulfils both of these requirements is often laborious and time-consuming. Moreover, allowance must be made for the fact that in the quantitative analysis of this kind of sample the relative volatilities (Section 1.3, Eq. 10), and consequently the concentrations of the compounds being analysed in the vapour phase, are largely dependent on the nature of the solvent. Essentially, the analysis is concerned with a system of at least the following seven components: (1) butyl acrylate, (2) butyl propionate, (3) styrene, (S) internal standard (*n*-butylbenzene), (4) DMF, (5) water and (6) dissolved polymer.

The volatilities of the substances being analysed relative to that of the internal standard in the actual mixture are defined as follows:

$$\alpha_{1S} = \frac{p_{01}}{p_{0S}} \left(\frac{\gamma_1}{\gamma_S}\right) 1\ 2\ 3\ \mathrm{S}\ 4\ 5\ 6$$

$$\alpha_{2S} = \frac{p_{02}}{p_{0S}} \left(\frac{\gamma_2}{\gamma_S}\right) 1\ 2\ 3\ \mathrm{S}\ 4\ 5\ 6$$

$$\alpha_{3S} = \frac{p_{03}}{p_{0S}} \left(\frac{\gamma_3}{\gamma_S}\right) 1\ 2\ 3\ \mathrm{S}\ 4\ 5\ 6$$

However, since compounds 1 2 3 5 6 and S are present in concentrations which are extremely low compared with that of 4, these equations may be written:

$$\alpha_{1S} = \frac{p_{01}}{p_{0S}} \left(\frac{\gamma_1}{\gamma_S}\right) 1\mathrm{S}\infty 4$$

$$\alpha_{2S} = \frac{p_{02}}{p_{0S}} \left(\frac{\gamma_2}{\gamma_S}\right) 2\mathrm{S}\infty 4$$

$$\alpha_{3S} = \frac{p_{03}}{p_{0S}} \left(\frac{\gamma_3}{\gamma_S}\right) 3\mathrm{S}\infty 4$$

i.e. in the actual analysis the relative volatilities are determined only by the dimethyl formamide. When this kind of effect is caused by the dissolved polymer (6) and the water (5), which are present in the ratio 1:1 in the dispersion being considered, it will be allowed for in the calibration. The concentration of water in the solvent is constant; however, the concentration of water in samples of the aqueous dispersion which have been weighed out will not be constant. Since it is extremely difficult to weigh out samples of aqueous dispersions in an accurately reproducible manner, the water content varies slightly from sample to sample. For measurements in the region from 3 to 3.6 weight % water shown in Fig. 36*

* Aqueous dispersions usually consist of about 50% polymer and 50% water. The weighed sample consists of 0.2 g dispersion in 3 cm^3 DMF. If 0.2 g samples of dispersion are weighed out, these will vary between 0.18 g and 0.22 g.

(vertical dashed lines) even in the most unfavourable case the relative error can only amount to 2%.

It follows from Fig. 36 that the sensitivity of this analysis can be enhanced by working at higher water concentrations, but precipitation of the dissolved polymer must then, of course, be taken into account. The corresponding effect due to the polymer must be examined in a similar manner to the above, although it has been established that it causes a relative error of only <1%.

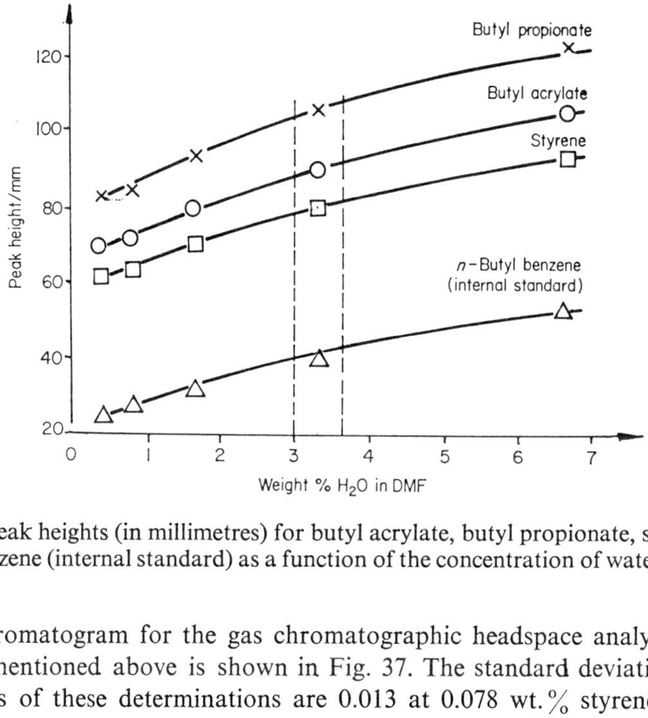

Fig. 36 Peak heights (in millimetres) for butyl acrylate, butyl propionate, styrene and *n*-butylbenzene (internal standard) as a function of the concentration of water in DMF.

The chromatogram for the gas chromatographic headspace analysis of the product mentioned above is shown in Fig. 37. The standard deviations (s) of the results of these determinations are 0.013 at 0.078 wt.% styrene; 0.07 at 0.63 wt.% butyl acrylate; and 0.015 at 0.077 wt.% butyl propionate.

The operating conditions for the chromatogram shown in Fig. 37 were as follows:

Instrument: Model F40 chromatograph (Perkin-Elmer, Bodenseewerk)
Column packing: Reoplex®, 15% on Chromosorb® DMCS
Column length: 5 m
Internal diameter of column: 2 mm
Carrier gas: N_2
Sampling time: 14 s
Column temperature: 120°C, isothermal
Sample temperature: 60°C
Recorder: 1 mV

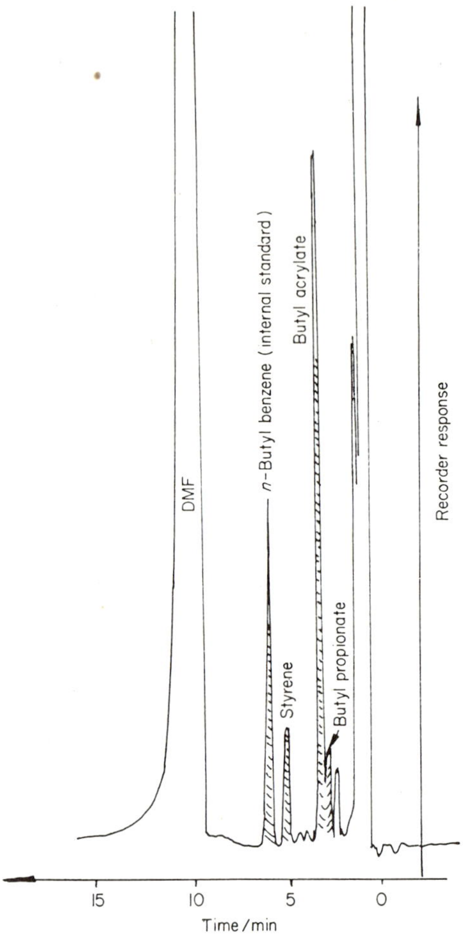

Fig. 37 Chromatogram obtained by gas chromatographic headspace analysis of an aqueous dispersion of a styrene–butyl acrylate copolymer.

THE DETERMINATION OF VINYL ACETATE IN THE POLYMER PRODUCED IN METHANOL SOLUTION[5]

The quantitative analysis of small amounts of vinyl acetate in the polymer produced in methanol solution demonstrates both the accuracy and the reproducibility of analysis by the headspace method. It is recommended that at the higher concentrations of vinyl acetate the samples should be suitably diluted with methanol before analysis (cf. Section 1.2). Analytical results obtained by the headspace method are compared in Table 10 with those obtained by the bromide–bromate titration of vinyl acetate in poly(vinyl acetate) prepared by polymerization in methanol solution.

TABLE 10

Determination of vinyl acetate by titration and headspace analysis of the polymer produced in methanol solution

Sample	*Wt. % vinyl acetate determined by titration*	*Wt. % vinyl acetate determined by headspace analysis* 1	2	3	4	*mean value*
A	7.0	7.6	7.7	7.5	7.6	7.6
B	0.5	0.7	0.7	0.7	—	0.7
C	19.0	19.1	18.7	18.0	—	18.6

1.632. Alcohol and toxic substances in blood

As in the case of aqueous dispersions of polymers, blood also comprises a solid and a liquid phase, with the difference that blood does not give a homogeneous solution in any solvent owing to its tendency to coagulate. The exact composition of the system water–blood sugar–fat–protein–globulin–albumin is of course not known accurately, but it is available in a pure form. The determination of blood alcohol has become very important as a result of the steady growth in road traffic. Established chemical oxidation methods in solution exist for this analysis; however, the other components present in blood, such as acetone or acetaldehyde are measured simultaneously. Direct gas chromatographic procedures for the determination of the concentration of alcohol in blood are also unsatisfactory, as is the case with aqueous dispersions of polymers, owing to the difficulties inherent in using syringes for the sampling of two-phase systems.[62]

On the basis of the originally published methods for the determination of blood alcohol levels,[63,64] Machata[65] and Jentsch[11] have developed the analysis to its present degree of accuracy by employing the headspace technique. The analysis of blood alcohol still constitutes the principal application of headspace analysis. A significant contribution to this field was the development of the fully automatic Automat Multifract F40 gas chromatograph (Perkin-Elmer, Bodenseewerk),[38] which gives results having standard deviations smaller than those which can arise from the uncertainties in the composition of blood. Standard ethanol solutions (Merck) are usually used instead of blood for calibrations, so that correction factors have to be included for blood and serum.

The following formula is used to calculate the results of the blood alcohol analysis:[38]

$$\%\ \text{ethanol} = \%\ \text{ethanol in standard sample} \cdot F \cdot \frac{h_e}{h_{st}}$$

where F – correction factor for blood (0.926) or serum (0.794); h_e – peak height for ethanol in the analysis; and h_{st} = peak height of the internal standard. *tert*-Butanol is usually employed as the internal standard, and since it has approximately the same polarity as ethanol, it produces similar changes in the activity coefficients.

Hauck and Terfloth[66] made a detailed study of the errors associated with this fully automatic determination of blood alcohol. They found, for example, that raising the bath temperature by 1°C produced a 5.5% increase in the peak heights for ethanol and the internal standard, although temperature had no influence on the ratio of the ethanol peak height to that of the internal standard. Variations in the peak height with the total amount of the sample were small, being about ±0.05%.

If a vacu-tube containing fluoride preservative was used to contain the blood sample, then the quantitative result of the analysis was influenced by the sodium fluoride dissolved in the blood. For example, for a sample in which the sodium fluoride concentration was 10 mg l^{-1} the ethanol peak height was 5% greater than that for a sample which did not contain sodium fluoride, although the ratio of the peak heights defined above remained constant. However, differences in the concentration of blood serum influence both the peak height and this ratio.

A number of other papers on the determination of blood alcohol by headspace analysis can only be mentioned briefly within the scope of this book.[67–70] For example, it has been reported that the syringe used to take the blood sample can also serve as the headspace analysis vessel if, when the equilibrium vapour pressure has been established, the vapour sample is removed by inverting the syringe and injecting into the gas chromatograph.[71] The concentration of ethanol can also be increased by adding salt, an example being the addition of 0.2 g sodium chloride to 0.2 cm^3 of blood.[72]

Basette *et al.* measured blood alcohol concentrations up to 0.3% with a standard error of 0.003–0.005%.[73] Glendenning and Harvey[74] described a method for the routine determination of blood alcohol, special features of which were the calibration technique, and the preparation of standard blood–alcohol solutions from ox blood. They also examined the influence of important analytical parameters such as concentration, temperature, sample size, salt concentration, etc. They observed an unexpected dependence of the peak height of ethanol on the sample size. Thus, a 0.1 cm^3 sample gave a peak height of 27.5 units, whereas a 4.0 cm^3 sample gave a peak height of 31.7 units.

Glendenning compared the method he developed for blood alcohol analysis with the autoclave method (oxidation with dichromate),[75] and the distillation method,[76] and found good agreement between them, as illustrated by the results given in Table 11.

Luckey[77] employed a special purpose gas chromatograph, the 'Alco-Analyzer', for the determination of blood alcohol. This instrument is equipped with a gas sampling loop which enables a reproducible amount of sample to be taken from the headspace. Serum bottles of 35 cm^3 capacity were used as headspace analysis vessels, and the sample size was 1 cm^3. Wilkinson, Wagner and Sedman[115] have developed a micromethod which requires only 20–50 μl of blood, by which it is possible to determine ethanol concentrations over the range 0.003–1.2 mg cm^{-3} with an accuracy of ±4.6%. The sample is taken into a special syringe fitted with a heated water jacket, and by maintaining the sample at 60°C instead

TABLE 11

Blood alcohol determination by three different methods
(from B. L. Glendenning *et al.*, *J. Forensic Sci.*, **14**, 136 (1969))

Sample	*Headspace analysis* % C_2H_5OH	*Autoclave method* % C_2H_5OH	*Distillation method* % C_2H_5OH
1	0.211	0.213	0.212
2	0.256	0.262	0.265
3	0.239	0.238	0.228
4	0.225	0.230	0.237
5	0.182	0.182	0.173
6	0.046	0.044	0.039
7	0.211	0.228	0.216
8	0.202	0.209	0.211
9	0.246	0.252	0.245
10	0.260	0.265	0.264

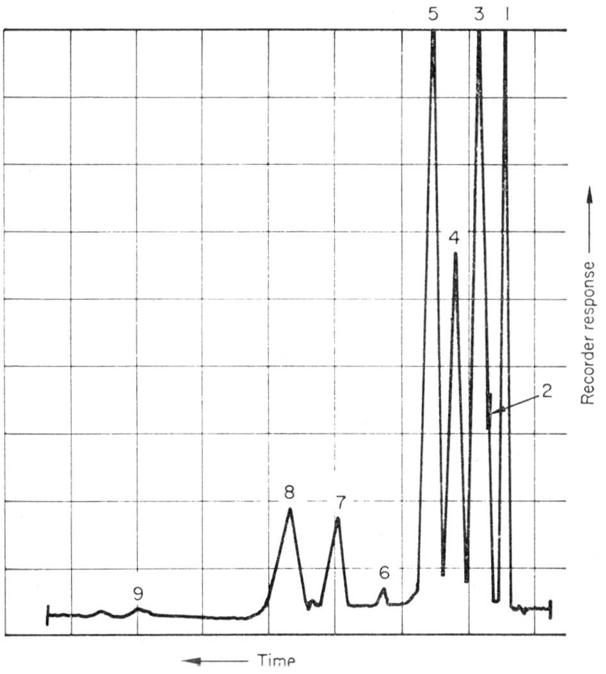

Fig. 38 Gas chromatographic headspace analysis of the volatile constituents in blood: 1, ether; 2, methanol; 3, acetone; 4, ethanol; 5, isopropanol; 6, carbon tetrachloride; 7, chloroform; 8, trichloroethylene; 9, acetaldehyde (from F. Rieders, *Gas Chromatographic Analysis in Toxicology: Theory and Application of Gas Chromatography in Industry and Medicine*, Grune and Stratton, New York (1968)).

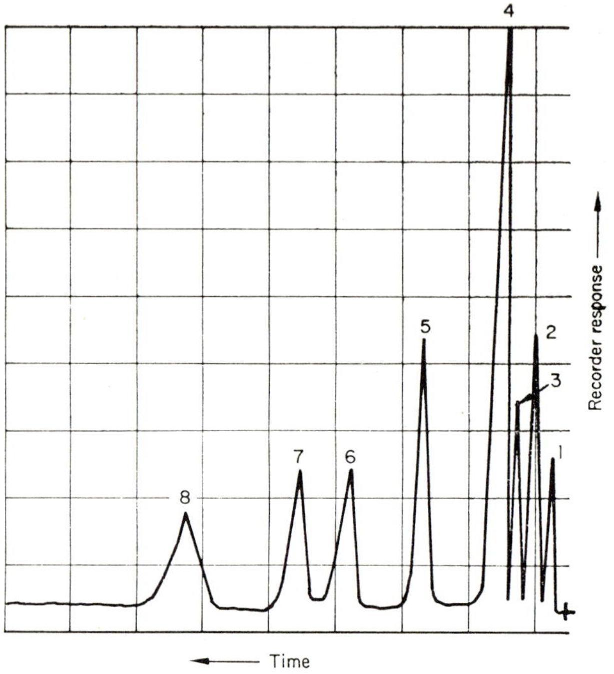

Fig. 39 The determination of volatile anaesthetics in blood by gas chromatographic headspace analysis: 1, ethylene; 2, cyclopropane; 3, vinyl methyl ether; 4, diethyl ether; 5, Halothan®; 6, chloroform; 7, trichloroethylene; 8, methoxyflurane (from F. Rieders, *Gas Chromatographic Analysis in Toxicology: Theory and Application of Gas Chromatography in Industry and Medicine*, Grune and Stratton, New York (1968)).

of the usual 30°C a fourfold increase in the concentration of alcohol is achieved. The addition of sodium nitrite to these samples inhibited the partial oxidation of the ethanol to acetaldehyde. Göke[117] has described a fully automatic and computerized technique for the determination of blood alcohol, and has discussed the qualitative and quantitative errors involved.

For toxicological purposes other compounds in addition to alcohol have to be determined in blood taken from corpses, and in blood samples which have been stored for long periods: these include acetaldehyde, methanol, propanol, butanol and amyl alcohol.[11,78] Other impurities which occur in blood and body fluids, such as ether,[79] and anaesthetics (Halothan®, chloroform, carbon tetrachloride, trichloroethylene and paraldehyde)[64] can be detected and measured by the headspace analysis method. A compilation of toxicologically important compounds and anaesthetics which have been analysed in blood was produced by Rieders.[80] Figure 38 shows a chromatogram of the headspace analysis of the toxicologically important constituents in blood, and Fig. 39 shows the corresponding trace for anaesthetics. Analytical conditions for the chromatograms in Figs. 38 and 39 were as follows.

Column: 6 feet, 1/8 inch, packed with 15 parts Flexol® 8N8, 10 parts diisodecyl phthalate, 3 parts PEG 600 on 42/60 mesh C22 firebrick
Column temperature: 110°C
Injection block: 120°C

Other examples which may be mentioned of this type of analysis of body fluids are the determination of acetone and β-ketobutyric acid in blood serum,[81] and methyl mercaptan in urine.[82]

1.633. Olfactory substances in milk

The headspace analysis of the substances which cause odours in milk has been chosen as the final example from this group. Like dispersions of blood, milk also consists of a solid and a liquid phase, i.e. 88 % water containing casein, albumin, fat, lactose, etc., and once again the medium can be obtained free from impurities. Consequently, Basette, Özeris and Whitnah[83] have been able to use headspace analysis to identify the substances which cause odours in milk. Sundararajan, Tobias and Whitney[84] have developed a method of analysing the headspace vapours over sterilized milk concentrate to enable changes in these olfactory substances to be determined during production and storage.

1.64. Group III—Solid Samples Which Dissolve Completely in a Solvent

Gas chromatographic headspace analysis is at present the most sensitive method for the direct determination of volatile constituents in solid materials, since it enables intricate and time-consuming concentration procedures to be avoided. Nevertheless, the direct quantitative headspace analysis of these materials is usually impossible, since it depends on the feasibility of calibrating the method.

The simplest cases arise when the solid is completely soluble in a liquid and the analysis can be carried out by the methods used for materials in Group I, or, for soluble aqueous dispersions, by the Group II methods. It is obvious that also in this group the influence of the dissolved solid on the relative volatilities of the components being analysed must be examined. Moreover, the search for an appropriate solvent again often involves problems with regard to its purity and chromatographic retention time. Another approach to the analysis of solid samples which should be mentioned here for the sake of completeness is to carry out the analysis of the solid sample directly by the headspace method and to perform the calibration separately. If an internal standard is added, this approach can provide a very accurate gas chromatographic determination of, for example, the concentration of styrene in polystyrene[85] by dissolving it in methylene chloride and precipitating with methanol, and the result of this analysis can be used to calibrate the headspace method. A calibration curve obtained by measurements of this kind is shown in Fig. 40, and it can be used as a basis for the quantitative headspace analysis of styrene in polystyrene.[5]

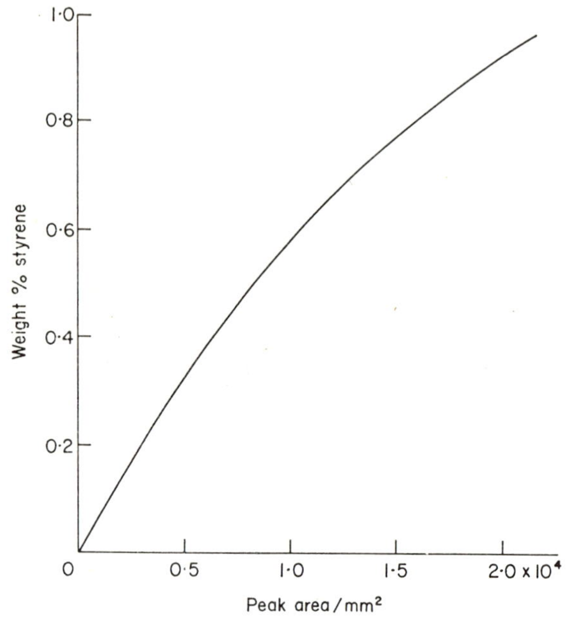

Fig. 40 Relationship between peak area and styrene concentration for the headspace analysis of polystyrene using the precipitation technique.

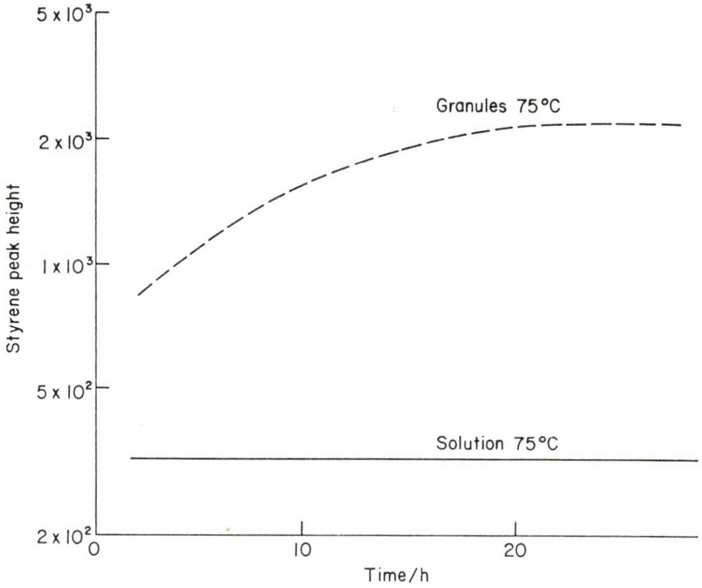

Fig. 41 Equilibration of styrene monomer between the gas phase and polystyrene granules and a polystyrene solution in DMF in the headspace analysis of polystyrene (from L. Rohrschneider, *Z. Anal. Chem.*, **255**, 345 (1971)).

This method has the unique advantage of requiring very little time, since the actual specimen can be analysed directly. However, as is seen in Fig. 40 the calibration graph obtained is not a straight line, but a curve, and this is observed with most solids. Furthermore, it takes considerably longer for the vapour pressure equilibrium to be established over solids than over solutions,* as Rohrschneider[86] has demonstrated for polystyrene. The approach to equilibrium between styrene and polystyrene granules is compared with the behaviour of a solution of the polymer in DMF in Fig. 41. It is therefore recommended that the headspace analysis of soluble solids be carried out with homogeneous solutions, and some examples of this approach are given in the following sections.

1.641. The determination of styrene and other volatile substances in polystyrene

Styrene and other volatile substances such as acrylonitrile, toluene, ethylbenzene, *o*-, *m*- and *p*-xylene, cumene, *n*-propylbenzene, butylbenzene and methylstyrene may be determined in polystyrene by the method developed by Rohrschneider,[86] in which a homogeneous solution of the polymer in DMF is employed. From 180 to 220 mg of polystyrene is weighed into the sample flask and 2 cm^3 of DMF, which already contains the internal standard (*n*-butylbenzene), is added. The polymer is dissolved by regularly agitating the mixture at 70°C. The calibration was carried out with five sample bottles each containing 2 cm^3 of a standard solution comprising a quantitatively known mixture in DMF of the compounds being determined. The standard solutions were sealed and equilibrated at 70°C. The gas chromatographic operating conditions for the analysis were as follows.

Gas chromatograph: Multifract F40
Column: 2 m Reoplex® 400 (Merck) 15% on Embacel® 60/100 mesh
Internal diameter of column: 3 mm
Column temperature: 100°C isothermal
Carrier gas: Nitrogen (9 cm s^{-1})
Duration of analysis: 16 min
Sampling time: 10 s
Injection block temperature: 100°C
Sample inlet manifold temperature: 90°C
Water thermostat temperature: 70°C
Detector: FID

Figure 42 shows the headspace chromatograms for the sample and a standard solution.

According to Rohrschneider, the limit of detection of styrene by this method

* Polymers having a glass transition point are an exception. If they are heated above the softening point, then equilibration occurs as rapidly as with solutions. The calibration can, of course, only be performed with pure monomers.[116]

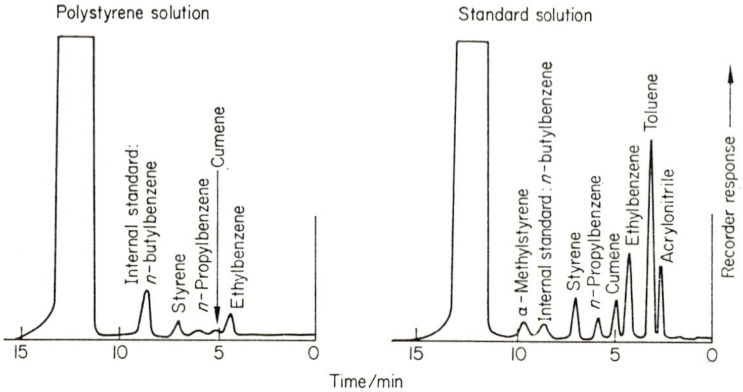

Fig. 42 Chromatogram of the gas phase above a polystyrene solution and above a standard mixture (from L. Rohrschneider, *Z. Anal. Chem.*, **255**, 345 (1971)).

is 10 mg per kg polystyrene. An analysis time of 6 min can be achieved by using a computer to calculate the results. The reproducibility of the method is so good that the calibration factors determined by Rohrschneider can be used on other Multifract F40 instruments with very little disparity. Thus, for example, the calibration factor reported for styrene of 0.3584 has been reproduced as 0.3500, and that of 0.0434 for acrylonitrile as 0.0430.[87] The advantage of headspace

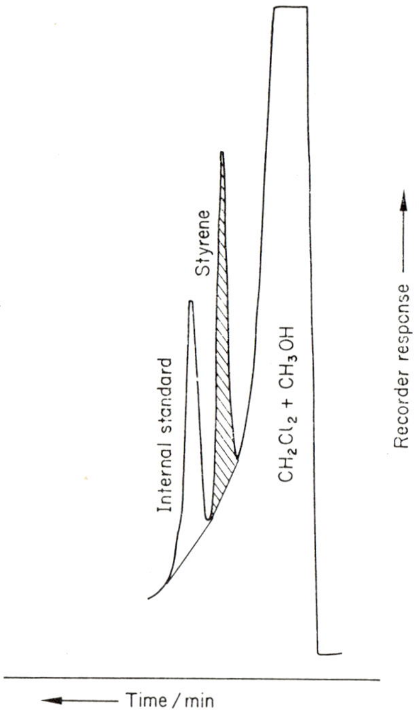

Fig. 43 Determination of styrene in polystyrene by the precipitation method.

analysis compared with the conventional reprecipitation method[85] can be seen on comparing the chromatogram shown in Fig. 42 with that in Fig. 43. The latter illustrates the conventional method of determining styrene in polystyrene by dissolving in methylene chloride and precipitating with methanol, followed by normal gas chromatographic sampling with a syringe. Except for the styrene peak, the only other identifiable peak is that due to the internal standard, since the solvent and precipitant are eluted together. The sample contains such a high proportion of these two constituents that other impurities cannot be determined at the same time, because the signal due to the solvent and the precipitant obscures the whole chromatogram. Furthermore, this procedure results in a low sensitivity for the detection of styrene.

The productivity, based on time and operator costs, is just as important as the accuracy for this kind of analysis, particularly when it is performed in industrial laboratories. The two methods are compared below from this point of view.

The essential operations involved, and the time and operator effort required for carrying out the determination of styrene in polystyrene by headspace analysis and by the precipitation method are as follows (assuming 30 samples).

Precipitation method	*Headspace analysis (Multifract F40 instrument and PEP1 computer—Perkin-Elmer Ltd.)*
Weighing of the polymer sample	Weighing of the polymer sample
Dissolution in methylene chloride	Dissolution in DMF
Precipitation with methanol	Introduction of sample into thermostat
Settling of the precipitate	Completely automatic performance of the analysis, including presentation of results
Manual gas chromatographic analysis; interpretation of results and preparation of report	
Time required: 16 h	Time required: 3 h
Staff required: 2	Staff required: 1

1.642. The determination of vinyl chloride in polyvinyl chloride

Puschmann[88] has shown that the vinyl chloride present in polyvinyl chloride can be determined in a similar manner to the styrene content of polystyrene. Of the two solvents which can be considered for this polymer, tetrahydrofuran (THF) and dimethylacetamide (DMA), the latter is preferable, since not only is its vapour pressure lower (shorter analysis time), but also it is generally available with a higher purity than THF. Diethyl ether is used as the internal standard, and is dissolved in the solvent before the polymer. The sample is equilibrated at a temperature of 50°C, which is adequate for the determination of less than 1 p.p.m. of vinyl chloride. The gas chromatographic conditions used are as follows.

Gas chromatograph: Multifract F40
Column: 4 m, 15% Polypropylene glycol 3000 on Kieselguhr 200–250 m^2 g^{-1} surface area (Camay, Switzerland)
Internal diameter of column: 4 mm
Column temperature: 60°C
Carrier gas: Nitrogen at 20 cm^3 min^{-1}
Detector: FID
Automatic operating conditions: Thermostat 50°C; sample inlet manifold 150°C; sampling time 5 s; duration of analysis 4 min; flushing period 0.5 min; interval between automatic cycles 0.5 min
Recording and evaluation of results: Hewlett Packard model 33 laboratory data handling system and 1 mV recorder
Sample preparation: 10% solution of PVC in DMA containing 100 p.p.m. diethyl ether as internal standard. Up to 30 samples can be introduced into the thermostatted sample holder at 50°C

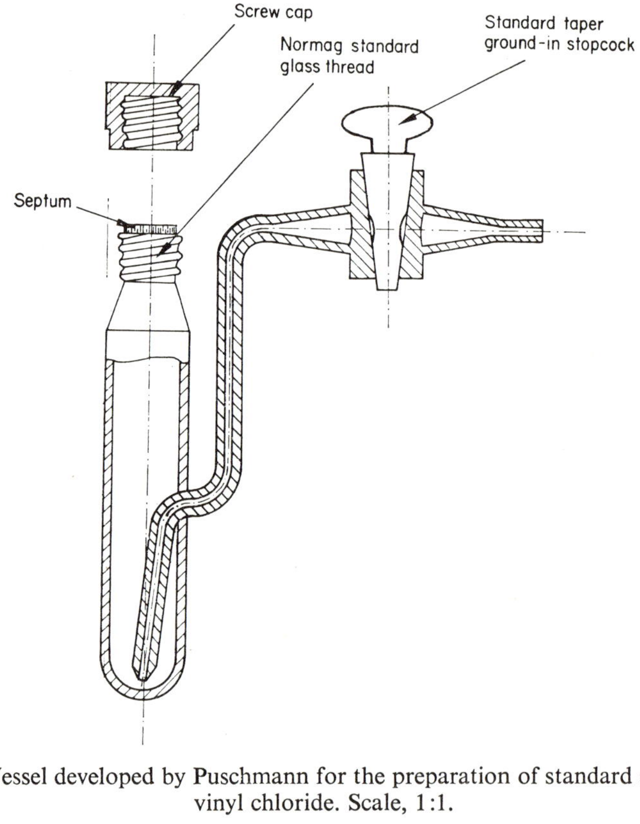

Fig. 44 Vessel developed by Puschmann for the preparation of standard solutions of vinyl chloride. Scale, 1:1.

When DMA is used as the solvent, after the first 15 analyses, each requiring a 6 min cycle, have been performed, the automatic sequence must be interrupted and the column temperature raised to 190°C to drive out the DMA. The column oven is then reset at 60°C and the automatic analysis sequence resumed for the remaining samples.

The standard solutions needed to calibrate this analysis are prepared in a vessel specially developed for the purpose, which is shown in Fig. 44. About 10 g of DMA is weighed with an accuracy of 1 mg into the glass vessel. Approximately 0.1 g of vinyl chloride is then introduced, with an accuracy of ±0.2 mg, from a gas cylinder via the gas inlet tube at such a rate that no indication of gas passing through the solution can be detected. Aliquots of this standard solution can be removed through the serum screw cap when required. The calibration factor is calculated in the normal manner from the equation

$$F_{VC} = \frac{\text{peak height (internal standard)} \cdot \text{p.p.m. by wt. (VC)}}{\text{p.p.m. by wt. (internal standard)} \cdot \text{peak height (VC)}}$$

where VC is vinyl chloride. Over the range 1–1000 p.p.m. by weight of vinyl chloride the calibration factor obtained in this way had a constant value of 0.73 with a standard deviation of 0.013. This factor was determined both in pure DMA and in a solution of PVC free from vinyl chloride, but no difference was observed between the two values, i.e. the relative partition coefficient is constant in these systems. The results of analyses are calculated from the expression:

$$\text{p.p.m. by wt. (VC)} = \frac{\text{peak height (VC)} \times F_{VC}\ \text{p.p.m. by wt. (internal standard)}}{\text{peak height (internal standard)}}$$

where F_{VC} is a calibration factor. The standard deviation (s) obtained in the analysis of PVC film was 0.06 at a vinyl chloride concentration of 3.7 p.p.m. and 0.47 for 27.8 p.p.m. of vinyl chloride. Chromatograms from the determination of 0.01 p.p.m. by weight of vinyl chloride and a corresponding blank estimation under identical conditions are shown in Fig. 45.

Puschmann used the same method to determine traces of vinyl chloride down to the level of about 0.03 p.p.m. by weight in olive oil (see also Section 1.624), and also in water-soluble food substitutes comprising water containing 3% acetic acid and 10% ethyl alcohol. He found a tenfold higher response for vinyl chloride compared with the determination in olive oil. Thain[89] used tetrahydrofuran as the solvent for the analysis of the same materials; however, in each of the analyses the solvent peak was incompletely resolved. The column packing used was 25% diisodecyl phthalate + 0.5% Atpet® 80 on Diatomite C® 60/72 mesh. The column was 2 m long and it was operated at 60°C. These analyses were also performed with a Multifract F40 chromatograph in which the bath temperature was set at 30°C and the injection time was 5 s. The problems which

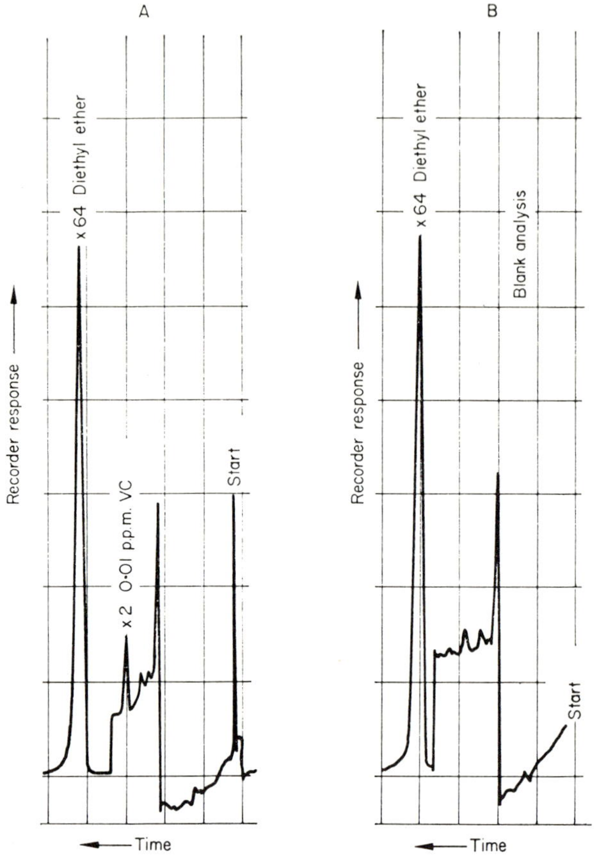

Fig. 45 Determination of vinyl chloride in polyvinyl chloride by the headspace analysis method (from H. Puschmann, *Angew. Makromol. Chem.*, **47**, 29 (1975)). See text for analytical conditions.

can arise from impurities in the solvent (THF) in these analyses are evident from the chromatograms reproduced in Fig. 46.

Dormehl[119] and Van Lierop[120] have also reported methods of determining traces of vinyl chloride down to the p.p.b. level in food substitutes. They recommended the Hall detector, which is specific for chlorine-containing compounds.

1.65. Group IV—Incompletely Soluble and Insoluble Solid Materials

If the solid material does not dissolve or is largely insoluble, it is unlikely that a quantitative headspace analysis can be performed, since neither can the main component be identified nor can the calibration standards be prepared. It may

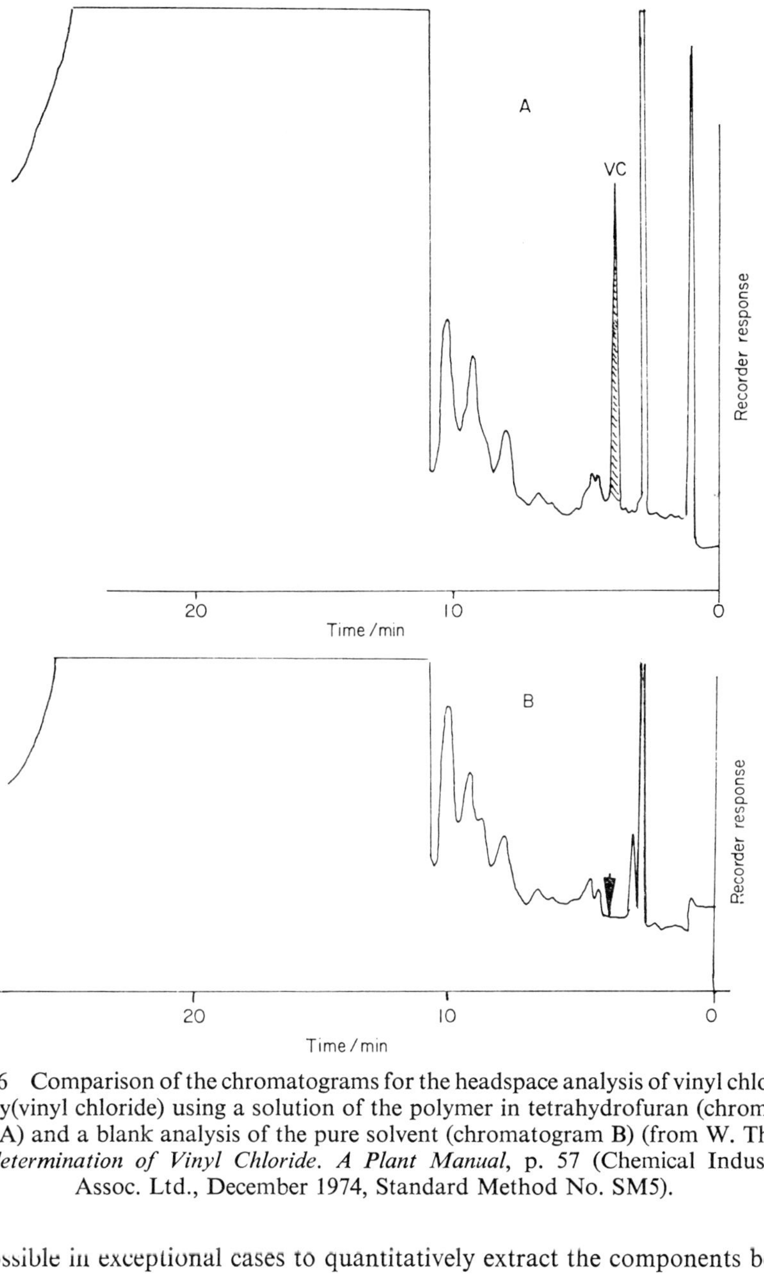

Fig. 46 Comparison of the chromatograms for the headspace analysis of vinyl chloride in poly(vinyl chloride) using a solution of the polymer in tetrahydrofuran (chromatogram A) and a blank analysis of the pure solvent (chromatogram B) (from W. Thain, *The determination of Vinyl Chloride. A Plant Manual*, p. 57 (Chemical Industries Assoc. Ltd., December 1974, Standard Method No. SM5).

be possible in exceptional cases to quantitatively extract the components being determined from the solid into a solvent and subsequently carry out a headspace analysis of this solution. However, in the majority of cases the gas chromatographic headspace analysis of materials in this group is limited to obtaining comparative chromatograms, although if they are produced under reproducible conditions, they can lead to valuable conclusions with regard to quality control, as will be seen from the examples discussed below.

1.651. Volatile substances in polyethylene

Since polyethylene is more or less insoluble in organic solvents, the low molecular weight impurities cannot be determined in solution, or by dissolving and reprecipitating the polymer. Moreover, solvent extraction techniques often involve loss of low-boiling components before they can be analysed. However,

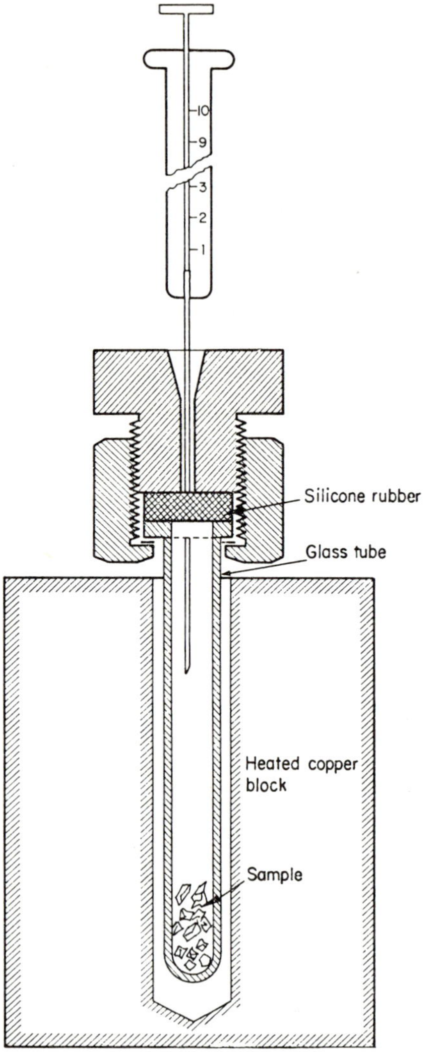

Fig. 47 Apparatus for the headspace analysis of solid materials (from T. R. Crompton and L. W. Myers, *Plastics and Polymers*, 205 (1968)).

a possible method of concentrating these volatile components in polyethylene consists of removing them from the polymer by volatilization at high temperatures (125–200°C). For this purpose Crompton and Myers[90] have developed a simple and very convenient device based on the principle of headspace analysis. The apparatus consists of a heated copper block containing a cylindrical glass vessel which holds the sample. The glass tube is sealed with a silicone rubber septum, and the sample can be withdrawn from the gas phase with a normal syringe and injected into the gas chromatograph (see Fig. 47).

Figure 48 shows gas chromatograms for the headspace vapours of polyethylene samples from different manufacturers.

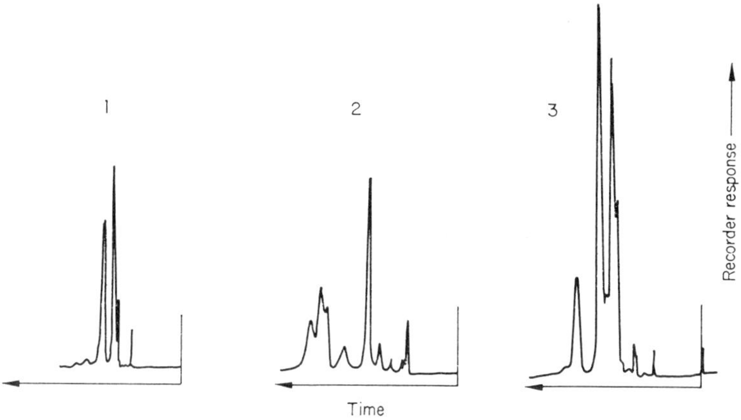

Fig. 48 Headspace chromatograms for various polyethylenes (1–3) (samples heated for period of 15 min at 125°C; column, 60 m copper capillary tube containing dibutyl phthalate; column temperature, 30°C) (from T. R. Crompton and L. W. Myers, *Plastics and Polymers*, 205 (1968)).

Although the extremely low concentrations in the vapour phase make it very difficult to identify the peaks in this analysis, even by combined gas chromatography and mass spectrometry, the method nevertheless provides a reproducible and rapid means of comparing and characterizing polyethylenes from different manufacturers on the basis of their volatile constituents. Figure 49 illustrates this kind of comparison between a polyethylene contaminated with compounds producing an odour (B) and an odourless material (A). It can be inferred from the two chromatograms that the components a, b, c and d are probably responsible for the odour.

It should be pointed out that comparisons of odours on the basis of peaks appearing on the chromatograms of the headspace vapours only very rarely lead to an unambiguous conclusion, since the sensitivity of the human nose is often very much higher than that of gas chromatographic detectors. It should also be remembered that very high boiling point polar constituents having an intense odour frequently do not produce a signal on the headspace chromatogram.

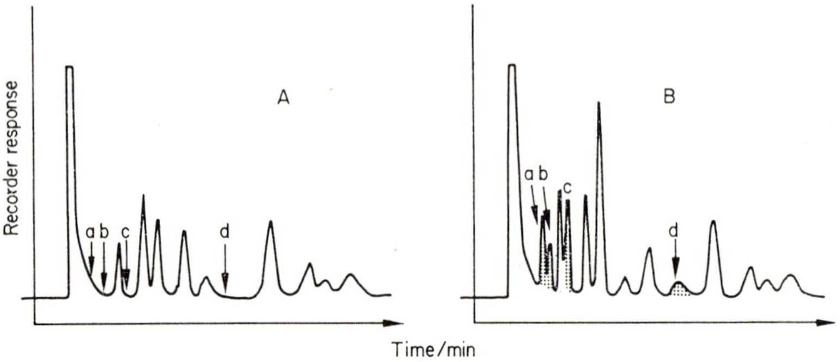

Fig. 49 Gas chromatograms of the headspace vapours from polyethylene contaminated with substances producing an odour (B) and an odourless sample (A) (for experimental conditions see Fig. 48).

This kind of investigation of polyethylene is of considerable interest, since polyethylene is used as a packaging material for milk. For example, gas chromatographic headspace analysis has shown the presence of ethane, propane, *n*-butane and but-1-ene, which are the secondary products of a free radical degradation of milk, and which can contribute to its 'plastics' taste.[91] Another example of this type of analysis is an investigation of guarana flavour added to materials in polyethylene containers.[92]

A new method of determining volatile compounds in insoluble polymers is discontinuous gas extraction, which, according to Kolb,[118] involves summation of the eluting peaks. The extraction follows an exponential law but it need not be exhaustive, since the residue can be determined by calculation.

1.652. Determination of the chemical stability of polymers

Gas chromatographic headspace analysis also provides a very simple and rapid method for testing the chemical stability of polymers.[93] Thus, for example, the resistance to water vapour, gaseous HCl, etc., may be studied at different temperatures in the actual sample container, since the vapours from it can be analysed directly.

Figures 50 and 51 illustrate this kind of investigation in the case of a polyacetal which has been exposed to gaseous HCl for half-an-hour.

1.653. The aroma and flavour constituents in natural products

The detection of aroma constituents in plants, fruit, tobacco, hops, mushrooms, etc., has been described in numerous papers. Thus, in an investigation of the aroma constituents in hops Narziss, Reicheneder and Forster[94] showed that it is only possible to perform a semi-quantitative analysis of the substances identified by measurement of their retention data. The attempted quantitative analysis of aroma constituents in powdered hops by reference to standard

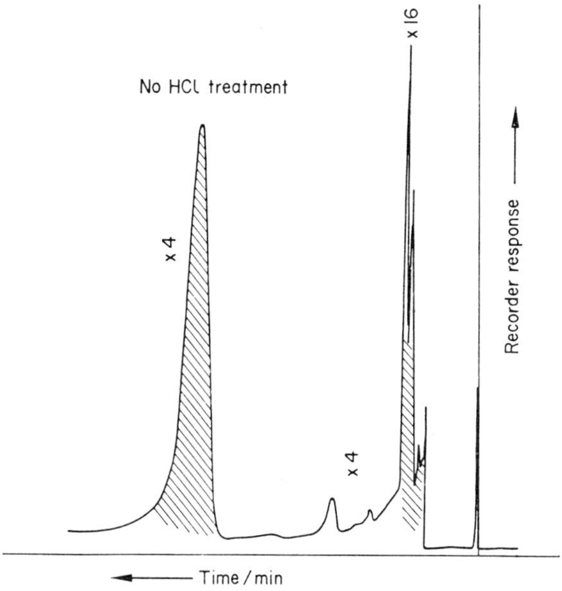

Fig. 50 Headspace analysis of a polyacetal before exposure to HCl.

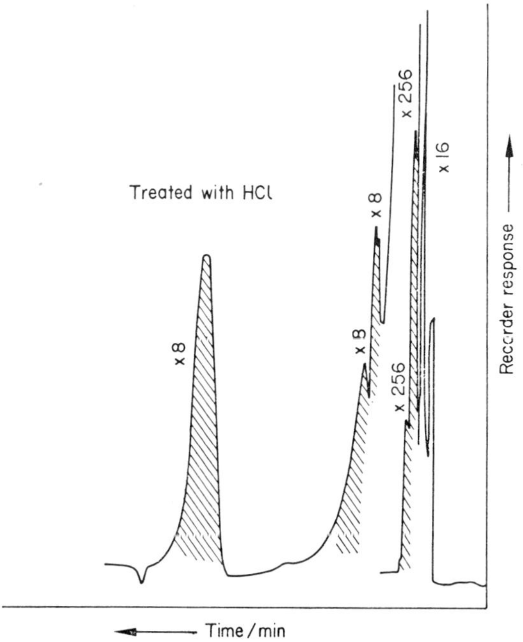

Fig. 51 Headspace analysis of a polyacetal after exposure to HCl.

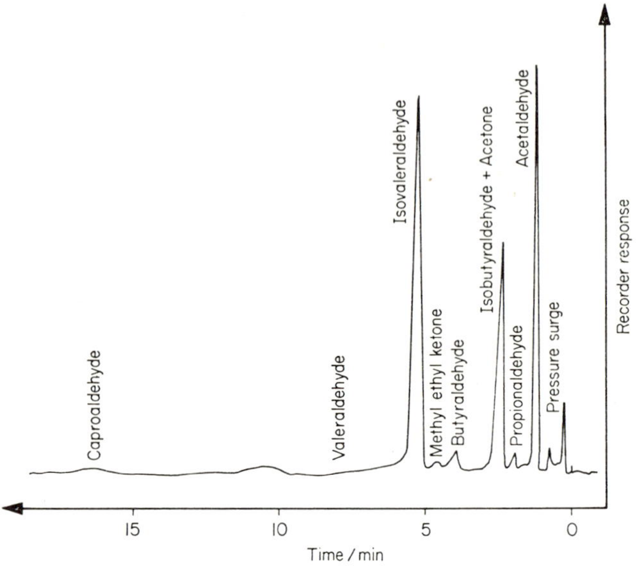

Fig. 52 Carbonyl compounds in barley (from B. Wagner, *Mschr. Brau.*, **24**, 285 (1971)).

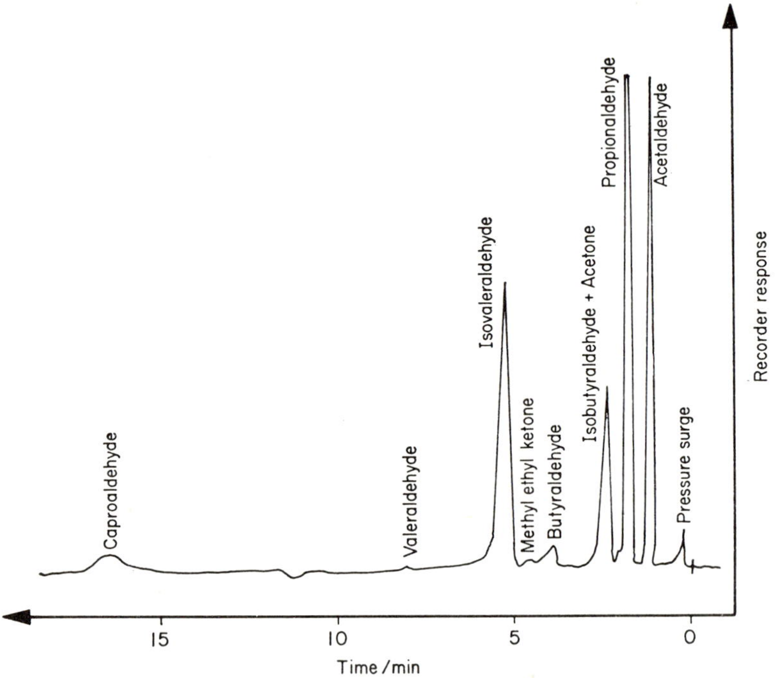

Fig. 53 Carbonyl compounds in malt (from B. Wagner, *Mschr. Brau.*, **24**, 285 (1971)).

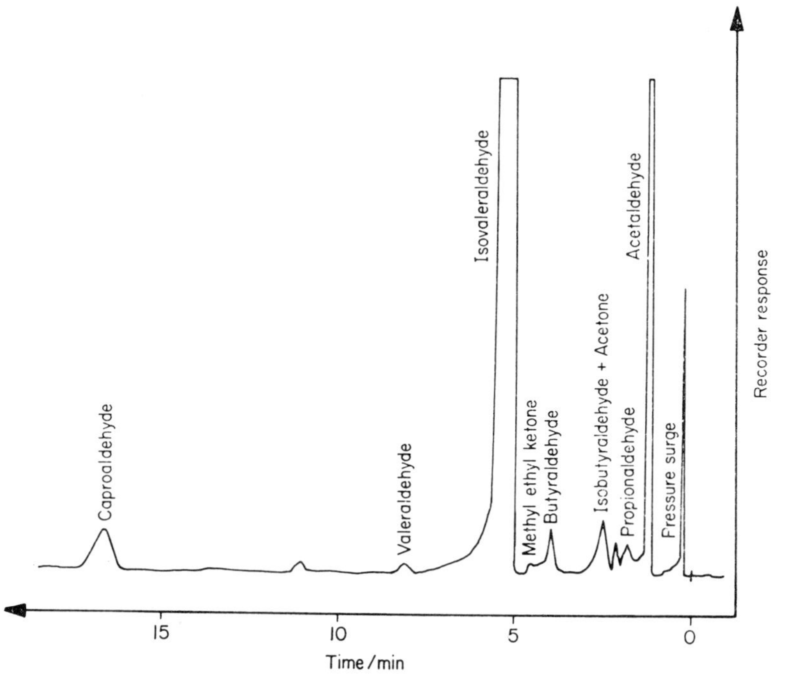

Fig. 54 Carbonyl compounds in green malt (from B. Wagner, *Mschr. Brau.*, **24**, 285 (1971)).

aqueous solutions has been unsuccessful, since the consequences of adsorption of the aroma substances in plant cellular material have not yet been established. The amount of each aroma constituent which passes into the gas phase from a standard solution can not be related to that from the plant material. Nevertheless, it is possible to compare different kinds of hops with each other by means of their headspace chromatograms, so that relative assessments of their qualitative and quantitative composition can be made.[95]

The volatile carbonyl compounds in barley, malt and green malt have been studied by Wagner.[96] Using a flame ionization detector, he was able to identify acetaldehyde, propionaldehyde, isobutyraldehyde, acetone, butyraldehyde, methylethylketone, isovaleraldehyde, valeraldehyde and caproaldehyde in barley, green malt and cured malt, as shown in the chromatograms in Figs. 52–54.

The operating conditions used to obtain the chromatograms in Figs. 52–54 were as follows.

Gas chromatograph: Multifract F40

Column: 2 m stainless steel, internal diameter 2.5 mm, 10% Carbowax 20M® and 10% Igepal® CO-880 on Chromosorb W-NAW® 60/80 mesh

Temperatures:
  Column oven: 60 or 90°C
  Injection block: 200°C
  Sampling needle: 150°C
  Water thermostat: 60°C
Gas flow rates:
  Carrier gas (N_2): 25 $cm^3\ min^{-1}$
  Synthetic air: 300 $cm^3\ min^{-1}$
  Hydrogen: 30 $cm^3\ min^{-1}$
Times:
  Equilibration time for samples in water thermostat: 20 min
  Injection: 3 or 6 s
  Duration of analysis: 19 min
  Flushing: 0.2 min
  Interval between analyses: 0.5 min

In order to identify the carbonyl compounds they were separated from the samples with nitrogen gas, condensed, and precipitated with 2,4-dinitrophenylhydrazine, and their retention times were compared with those of the corresponding pure substances. A 4 m Reoplex® 400 column was used in addition to the column mentioned above.

During the quantitative analyses involved in the above investigations, it was found that the water present in the samples had a considerable influence on the concentration of the carbonyl compounds in the headspace. Therefore, no absolutely quantitative values could be obtained for the amounts of these substances in barley, green malt or malt. Nevertheless, it was established that propionaldehyde is present in considerably higher concentrations in malt than in barley or green malt. Other similar investigations have been concerned with the characterization of the substances producing odours and aromas from cultured mushrooms[25] and fruits.[97]

In a study of the aroma from apples Drawert *et al.*[26] compared the headspace method with liquid–liquid extraction. They obtained different results from the two methods, and concluded that headspace analysis is undoubtedly very suitable for the determination of the composition of the aromatic substances in a volume of gas at a particular time, but that compared with the extraction method it does not give reliable quantitative results.

To provide a better comparison between the results of the gas chromatographic headspace method and the extraction method Romani and Ku[98] determined the peak area for each component for kilogram amounts of fruit. A 5 cm^3 sample was taken from the gas phase above 1.5 kg of fruit. Although an accurate value for the absolute concentrations in the fruit, of the substances producing its aroma, cannot be obtained from the instantaneous composition of the vapour, conclusions can be drawn regarding the approximate composition of the aroma.[99] The identification of substances producing the aromas from potatoes, carrots and pears stored for varying periods was reported by Buttery and Teranishi,[100] and in greater detail by Gascó, Barrera and de la Cruz.[101] The latter proposed the term 'aromagram' for the headspace fingerprint chromatograms recorded in such investigations, and it is certainly an apt method of referring to them.

The enzymatic formation of volatile compounds in raspberries has been characterized by Weurman.[102] Volatile sulfur compounds can be detected by the same method,[103] an example being the formation of methyl sulfide, which has been detected in tinned tomatoes. This was done by first absorbing the vapour in concentrated sulfuric acid, diluting, and then analysing the headspace above the solution. From 1.6 to 7.9 p.p.m. of methyl sulfide could be determined in this way. At 100°C and atmospheric pressure between 2 and 6 p.p.m. of this substance is formed.[104]

In Turkish tobacco as many as 35 components have been detected and identified by means of functional group reactions of aldehydes, ketones and esters. These reactions were carried out in the headspace analysis vessel using acidified or alkaline hydroxylamine solution, mercuric chloride and potassium permanganate.[105]

The combination of functional group analysis and headspace analysis has also been employed to identify the volatile substances from dairy products, such as kefir cheese and sheep-milk cheese.[16] Another field in which the method has been applied is the study of flavouring agents commonly employed in pharmacy, such as, for example, those used to give medicines a raspberry or banana flavour.[106]

Mackay, Lang and Berdick[107] investigated the volatile constituents in the headspace vapours over peppermint oil, bananas, roasted coffee, brandy, whisky, etc., and this enabled them to simply and rapidly establish the quality and genuineness of these products on the basis of their aromas, without having to carry out precise qualitative and quantitative analyses. Figure 55 shows this kind of comparison between freshly sliced banana (chromatogram A) and an artificial banana flavouring (chromatogram B).

Gas chromatographic headspace analysis is frequently employed in the determination of gaseous and volatile substances in tinned foods. For example, tins of apple sauce have been reported to contain 18.8 vol.% CO_2, 18.4 vol.% H_2, 1.7 vol.% O_2, 61 vol.% N_2 and 0.13 vol.% CO.[108] NO and N_2O were also detected, and they are the gaseous products of the reduction of nitro-compounds.[109] Special devices which can remove samples from tins, such as the Headspace Gas Sampler manufactured by Beckman, or home-made samplers like that described by Hurst,[110] are employed in this type of analysis.

The determination of residual solvents such as acetone, *n*-hexane, *n*-heptane and benzene in oleoresins is also a very important analysis.[111] Heins *et al.*[18] used capillary columns in a combined gas chromatograph and mass spectrometer to determine the composition of the headspace materials over tea and coffee beans, and were able to identify 24 compounds from tea and 60 compounds from

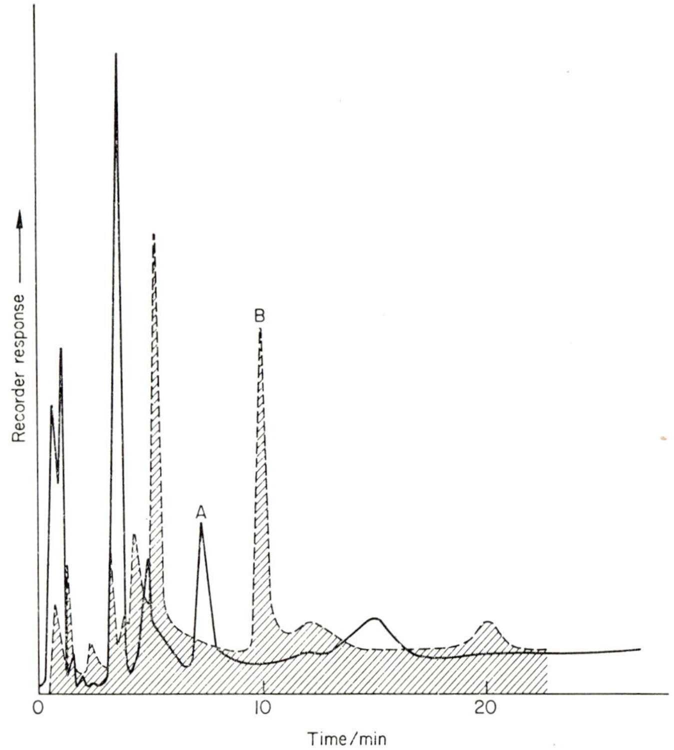

Fig. 55 Comparison of the headspace chromatograms from sliced banana (A) and banana flavouring (B) (from D. A. M. Mackay *et al.*, *Anal. Chem.*, **33**, 1369 (1961)).

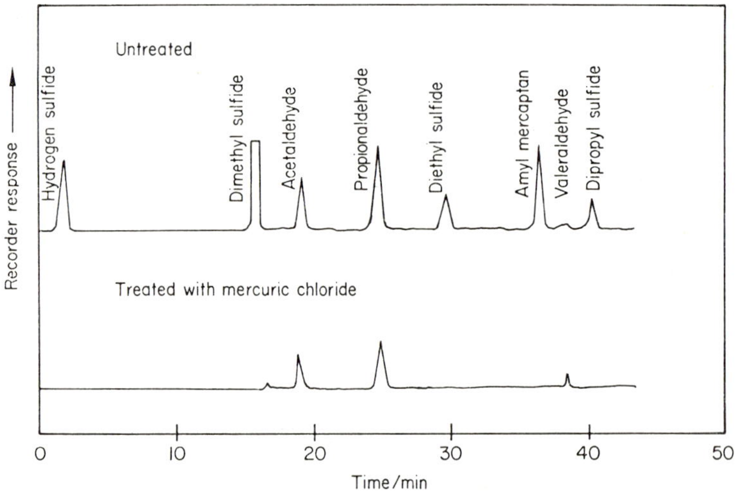

Fig. 56 Elimination of sulfides from aqueous extracts of clam meat by reaction with $HgCl_2$ (from J. M. Mendelsohn and R. O. Brooke, *Food Technol.*, **22**, 1162 (1968)).

coffee. Mendelsohn and Brooke[112] examined the headspace vapours produced during the storage of clam meat, and in addition to a number of unsaturated hydrocarbons found aldehydes, ketones and sulfur compounds. Dimethyl sulfide is the predominant headspace component, and is believed to be responsible for the characteristic odour of clams. The sulfides were identified by causing a portion of the headspace gas from the concentrated aqueous extract of the clam meat to react with $HgCl_2$, and performing comparative headspace analyses (Fig. 56).

Buttery *et al.*[113] in an investigation of volatile constituents of foodstuffs found that the relative volatilities of these substances can be satisfactorily simulated in water if water itself is the main constituent of the foodstuff concerned.

1.7. FUTURE PROSPECTS FOR HEADSPACE ANALYSIS

The examples of analytical investigations given above are not claimed to be exhaustive; they merely demonstrate the wide range of existing applications of gas chromatographic headspace analysis and indicate possible future uses for this method.

If, for example, it were necessary to quantitatively determine substances having a low vapour pressure, such as the high-boiling materials which cause the odours from polymers, this could only be achieved if the thermostatic baths in the headspace analysis instruments could be used with liquids of higher boiling point than water. This in turn necessitates the use of headspace vessels which can withstand high pressures and which incorporate an appropriately designed sampling port, since a rubber septum would no longer constitute an acceptable seal.

The flame ionization detector has been used in most instances because of the low concentrations of materials in the headspace, although it is known that it does not respond to all vapours and gaseous compounds. This precludes the possibility of determining traces of inert and inorganic gases, and of formaldehyde, formic acid and, especially, water.

As far as the determination of water in solids and liquids is concerned, Kolb[114] has recently demonstrated some interesting possibilities by using a sensitive thermal conductivity detector in association with a Porapak S® column. This technique may be used to determine the concentration of water in, for example, candy, which is dissolved in DMF for this purpose. Figure 57 shows a measurement carried out in this way by Kolb. The thermal conductivity detector is also to be preferred to the flame ionization detector when there are high concentrations of materials present in the headspace, since there is then the risk that the FID response will be non-linear.

The operating conditions used to obtain the chromatograms in Fig. 57 were as follows.

Gas chromatograph: Multifract F42 fitted with thermal conductivity detector (Perkin-Elmer, Bodenseewerk)
Column: Porapak S, 1/8 in × 2 m stainless steel
Column temperature: 160°C isothermal

Carrier gas: He at 25 $cm^3\ min^{-1}$
Water bath: 65°C
Sampling time: 5 s
Injection port and detector temperature: 200°C
Attenuation: ×16

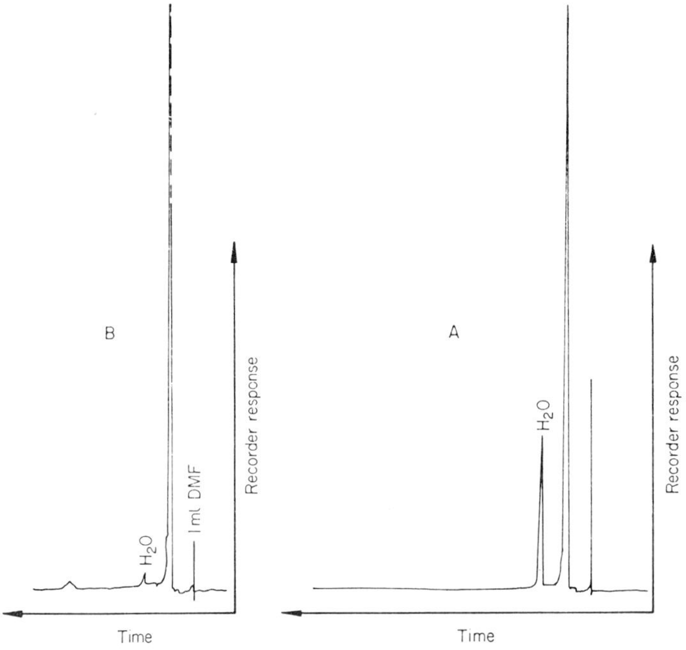

Fig. 57 Determination of water in candy by headspace analysis (from B. Kolb, private communication, Perkin-Elmer, Bodenseewerk). Trace A, actual analysis; trace B, blank analysis.

References

1. HACHENBERG, H., *Industrial Gas Chromatographic Trace Analysis*, Heyden and Son Ltd., London (1973).
2. BEREZKIN, V. G. and TATARINSKII, V. S., *Gas Chromatographic Analysis of Trace Impurities*, Consultants Bureau, New York (1973) (English edition).
3. HÁLA, E., PICK, J., FRIED, V. and VILIM, O., *Gleichgewicht Flüssig-Dampf* Akademie Verlag, Berlin (1960).
4. KOLB, B., *CZ—Chemie-Technik*, **1**, 87 (1972).
5. HACHENBERG, H., *TIPS* 41, *GC*, Bodenseewerk Perkin-Elmer, April (1970).
6. KORTÜM, G., *Lehrbuch der Elektrochemie*, p. 191, Verlag Chemie GmbH Weinheim Bergstrasse (1957).
7. EGGERT, J., *Lehrbuch der Physikalischen Chemie*, p. 633, S. Hirzel Verlag, Stuttgart (1960).
8. MACHATA, G., *Blutalkohol*, **4**, 252 (1967).

9. JENNINGS, W. G., *J. Food Sci.*, **30**, 445 (1965).
10. HACHENBERG, H., unpublished work.
11. JENTZSCH, D., KRÜGER, H. and LEBRECHT, G., *Angewandte Gaschromatographie*, Part 10, Bodenseewerk Perkin-Elmer (1967).
12. HACHENBERG, H., unpublished work.
13. Ref. 1, p. 36.
14. Ref. 1, p. 9.
15. BASETTE, R., ÖZERIS, S., and WHITBAH, C. H., *Anal. Chem.*, **34**, 1540 (1962).
16. PALO, V., *Chromatographia*, **4**, 55 (1971).
17. Ref. 1, p. 53.
18. HEINS, J. TH., MAARSE, H., TEN NOEVER DE BRAUW, M. D. and WEURMAN, C., *J. Gas Chromatog.*, 395 (1966).
19. LIEBICH, H. M., KÖNIG, W. A. and BAYER, E., *J. Chromatog. Sci.*, **8**, 527 (1970).
20. ANGELINI, P., BAZINET, M. L. and MERRITT, C., Jr., *160th ACS Nat. Meeting*, Chicago, Sept. 14–18 (1970).
21. HORN, O., SCHWENK, O. and HACHENBERG, H., *Brennstoff-Chem.*, **39**, 336 (1958).
22. PIORR, W., *Glas-Instr-Tech.*, **12**, 175 (1968).
23. BINDER, H., *Z. Anal. Chem.*, **244**, 353 (1969).
24. NOVÁK, J., VASÁK, V. and JANAK, J., *Anal. Chem.*, **37**, 660 (1965).
25. SPRECHER, E. and STRACKENBROCK, K. H., *Z. Naturf.*, **18b**, 495 (1963).
26. DRAWERT, F., HEIMANN, W., EMBERGER, R. and TRESSL, R.. *Chromatographia*, **2**, 77 (1969).
27. GOTTAUF, M., *Z. Anal. Chem.*, **218**, 175 (1966).
28. HORNSTEIN, I. and CROWE, P. R., *Anal. Chem.*, **34**, 1355 (1962).
29. MORGAN, M. E. and DAY, E. A., *J. Dairy Sci.*, **48**, 1382 (1965).
30. BINDER, H., *J. Chromatog.*, **25**, 189 (1966).
31. NAWAR, W. W. and FAGERSON, I. S., *Food Technol.*, **16**, 107 (1962).
32. ANDERSSON, J. and von SYDOW, E. *160th Nat. Meeting ACS*, Chicago, I, 11 Sept. Progr. Abstr. No. AGFD-5 (1970).
33. DRAVNIEKS, A. and O'DONNELL, A., *160th Nat. Meeting ACS*, Chicago, I, 11 Sept. Progr. Abstr. No. AGFD-2 (1970).
34. VITENBERG, A. G., IOFFE, B. V. and BORISOV, V. N., *Chromatographia*, **7**, 610 (1974).
35. CULHAM, A. D., *Brauwissenschaft*, **22**, 257 (1969).
36. RONKAINEN, P., *Kem. Teollismus*, **26**, 215 (1969).
37. ZENZ, H. and KLAUSHOFER, H., *Mitt. Vers. Anst. Gär. Gew.*, **22**, 175 (1968).
38. JENTZSCH, D., KRÜGER, H., LEBRECHT, G., DENCKS, G. and GUT, J., *Z. Anal. Chem.*, **236**, 112 (1968).
39. PAUSCHMANN, H., *Chromatographia*, **3**, 376 (1970).
40. GÖKE, G., *Chromatographia*, **5**, 622 (1972).
41. RATKOVICS, F., *Acta. Chim. Acad. Sci. Hung.*, **49**, 71 (1966).
42. MIETHKE, H., *Deut. Lebensm.-Rdsch.*, **65**, 379 (1969).
43. MAIER, H. G., *J. Chromatog.*, **50**, 329 (1970).
44. DAVIS, P. L., *J. Chromatog. Sci.*, **8**, 423 (1970).
45. ÖZERIS, S. and BASETTE, R., *Anal. Chem.*, **35**, 1091 (1963).
46. KEPNER, R. E., MAARSE, H. and STRATING, J., *Anal. Chem.*, **36**, 77 (1964).
47. JENTZSCH, D., KRÜGER, H. and LEBRECHT, G., *Angewandte Gaschromatographie*, Bodenseewerk Perkin-Elmer, Vol. 19, Part 9 (1967).
48. CORCORAN, E. F., CORWIN, J. F. and SEBA, D. B., *J. Am. Water Works Assoc.*, **59**, 782 (1967).
49 PLODER, I., *Deut. Lebensm.-Rdsch.*, **70**, 401 (1974).

50. MÄNDL, B., WULLINGER, F., BINDER, W. and PIENDL, A., *Brauwissenschaft*, **22**, 477 (1969).
51. JAHNSEN, V. J. and HORN, G. B., *Am. Soc. Brew. Chem.*, 194 (1965).
52. ZENZ, H. and KLAUSHOFER, H., *Mitt. Vers. Gär. Gew.*, **22**, 175 (1968).
53. BARON, G. and WAGNER, B., *Mschr. Brau.*, **24**, 123 (1971).
54. WAGNER, B. and BARON, G., *Mschr. Brau.*, **24**, 225 (1971).
55. OLSEN, A., *Wallerstein Labs. Commun.*, **33**, 19 (1970).
56. MAULE, D. R., *J. Inst. Brew.*, **73**, 351 (1967).
57. GROB, K. and GROB, G., *6th Int. Symp. Advan. Chromatography*, Miami Beach, June (1970).
58. DAVIS, P. L., *Hort. Sci.*, **4**, 117 (1969).
59. KOLB, B., WIEDEKING, E. and KEMPEN, B., *Angewandte Chromatographie*, Part 11-11E, Bodenseewerk Perkin-Elmer (1968).
60. HACHENBERG, H., *Angewandte Chromatographie*, Part 25, Bodenseewerk Perkin-Elmer (1975).
61. HENNEBERG, D. and SCHOMBURG, G., *Z. Anal. Chem.*, **215**, 424 (1966).
62. GLENDENNING, B. L. and HARVEY, R. A., *J. Forensic Sci.*, **14**, 136 (1969).
63. GOLDBAUM, L., in *Progress in Chemical Toxicology, Vol.* 1 (ed. A. Stohnan), Academic Press, New York (1963).
64. CURRY, A. S., HURST, G., KENT, N. R. and POWELL, H., *Nature*, **195**, 603 (1962).
65. MACHATA, G., *Microchem. Acta*, 262 (1964).
66. HAUCK, G. and TERFLOTH, H. P., *Chromatographia*, **2**, 309 (1969).
67. SAVORY, J., SUNDERMANN, F. W., ROSZEL, N. O. and MUSHAK, P., *Clin. Chem.*, **14**, 132 (1968).
68. COLDWELL, B. B., SALOMONRAJ, G., TRENHOLM, H. L. and WIBERG, G. S., *Clin. Toxicol.*, **4**, 99 (1971).
69. DURITZ, G. and TRUITT, E. B., *Q. J. Studies Alcohol*, **25**, 498 (1965).
70. NATELSON, S. and STELLATE, R. L., *Microchem. J.*, **9**, 245 (1965).
71. GOLDBAUM, L. R., DOMANSKI, T. J. and SCHLOEGEL, L. J., *J. Forensic Sci.*, **9**, 63 (1964).
72. WALLACE, J. E. and DAHL, E. V., *Am. J. Clin. Pathol.*, **48**, 152 (1966).
73. BASETTE, R. and GLENDENNING, B. L., *Microchem. J.*, **13**, 374 (1968).
74. GLENDENNING, B. L. and HARVEY, R. A., *J. Forensic Sci.*, **14**, 136 (1969).
75. KINSLEY, G. R. and CURRENT, H., *J. Lab. Clin. Med.*, **35**, 294 (1950).
76. HEISE, H. A., *Am. J. Clin. Pathol.*, **4**, 182 (1934).
77. LUCKEY, M. L., *J. Forensic Sci.*, **16**, 120 (1971).
78. PETRANECK, G., *Wiss. Z. Humboldt-Univ. Berlin*, 668 (1966).
79. Ref. 11, p. 3.
80. RIEDERS, F., *Gas Chromatographic Analysis in Toxicology: Theory and Application of Gas Chromatography in Industry and Medicine* (ed. H. S. Kroman and S. R. Bender), p. 266, Grune and Stratton, New York (1968).
81. van STREKELENBURG, G. J. and DE BRUYN, J. W., *Clin. Chim. Acta*, **28**, 233 (1970).
82. FIELD, T. G. Jr. and GILBERT, J. B., *Anal. Chem.*, **38**, 628 (1966).
83. BASETTE, R., ÖZERIS, S. and WHITNAH, C. H., *J. Food Sci.*, **28**, 84 (1962).
84. SUNDARARAJAN, N. R., TOBIAS, J. and WHITNEY, R., *J. Dairy Sci.*, **51**, 1169.
85. PFAB, W. and NOFFZ, D., *Z. Anal. Chem.*, **195**, 37 (1963).
86. ROHRSCHNEIDER, L., *Z. Anal. Chem.*, **255**, 345 (1971).
87. HACHENBERG, H., unpublished work.
88. PUSCHMANN, H., *Angew. Makromol. Chem.*, **47**, 29 (1975).

89. THAIN, W., *The Determination of Vinyl Chloride. A Plant Manual*, p. 57 (Chemical Industries Assoc. Ltd., December 1974, Standard Method No. SM5).
90. CROMPTON, T. R. and MYERS, L. W., *Plastics Polymers*, 205 (1968).
91. KIERMEIER, F. and STROH, A., *Z. Lebensm.-Untersuch-Forsch.*, **141**, 216 (1969).
92. DONATO, S. J., *J. Pharmacol. Sci.*, **56**, 759 (1967).
93. HACHENBERG, H., unpublished work.
94. NARZISS, L., REICHENEDER, E. and FORSTER, A., *Brauwissenschaft*, **23**, 289 (1970).
95. HAUTKE, P. and PETRICEK, D., *Wallerstein Labs. Commun.*, **33**, 89 (1970).
96. WAGNER, B., *Mschr. Brau.*, **24**, 285 (1971).
97. TERANISHI, R., BUTTERY, R. G. and LUNDIN, R. E., *Anal. Chem.*, **34**, 1033 (1962).
98. ROMANI, R. J. and KU, L., *J. Food Sci.*, **31**, 558 (1966).
99. BROWN, D. S., BUCHANAN, J. R. and HICKS, J. R., *Proc. Am. Soc. Hort. Sci.*, **88**, 98 (1966).
100. BUTTERY, R. G. and TERANISHI, R., *Anal. Chem.*, **33**, 1439 (1961).
101. GASCÓ, L., BARRERA, R. and de la CRUZ, F., *J. Chromatog. Sci.*, **7**, 228 (1969).
102. WEURMAN, C., *Food Technol.*, 531 (1961).
103. BAILEY, S. D., BAZINET, M. L., DRISCOLL, J. C. and McCARTHY, A. I., *J. Food Sci.*, **26**, 2 (1961).
104. MIERS, J. C., *J. Agric. Food Chem.*, **14**, 149 (1966).
105. SWAIN, A. P., PETERSON, R. F. and STEDMAN, R. L., *J. Sci. Food Agric.*, **17**, 349 (1966).
106. WELBY, J., HAUTBOUT, F., LUONG, T. T., PUISIEUX, F. and LEHIR, A., *Ann. Pharm. Franç.*, **28**, 623 (1970).
107. MACKAY, D. A. M., LANG, D. A. and BERDICK, M., *Anal. Chem.*, **33**, 1369 (1961).
108. LOPEZ, A. and KREBS, B. S., *Food Technol.*, **21**, 477 (1967).
109. ELKINS, E. R., KIM, E. S. and FARROW, R. P., *Food Technol.*, **23**, 1419 (1969).
110. HURST, R. E., *Anal. Chem.*, **47**, 1221 (1975).
111. LABRUYÈRE, B., OLSTHOORN-DE LEEUW, C. and SMEENGE, F., *Perfumery Essent. Oil Record*, **59**, 206 (1968).
112. MENDELSOHN, J. M. and BROOKE, R. O., *Food Technol.*, **22**, 1162 (1968).
113. BUTTERY, R. G., BOMBEN, J. J., GUADAGINI, D. G. and LING, L. C., 160*th ACS Nat. Meeting*, Chicago, September 14–18 (1970).
114. KOLB, B., private communication, Perkin-Elmer, Bodenseewerk.
115. WILKINSON, P. K., WAGNER, J. G. and SEDMAN, A. J., *Anal. Chem.*, **47**, 1506 (1975).
116. SEBESTYEN, N., private communication, Perkin-Elmer Corp., Norwalk, USA (1975).
117. GÖKE, G., *Blutalkohol*, **10**, 281 (1973).
118. KOLB, B., *Colloquim über die GC-Dampfraumanalyse*, Bodenseewerk Perkin-Elmer, Überlingen, BRD, 20/21 October (1975).
119. DORMEHL, M., Ref., 118.
120. VAN LIEROP, J. B., Ref. 118.
121. BENCSATH, F. A., Ref. 118.

PART 2

Physico-Chemical Applications

2.1 INTRODUCTION

If a system containing vapour and the corresponding condensed phase is allowed to reach equilibrium, and the vapour is analysed by the gas chromatographic headspace method, then this combination is an experimental technique which is of interest not only to analysts, but also, of course, to thermodynamicists. In fact this procedure was simultaneously, but completely independently, developed, on the one hand by analysts for chemical determinations, and on the other hand by thermodynamicists for the study of phase equilibria. The equipment designed for analytical purposes is largely automatic, and mostly available commercially, and there have recently been attempts to utilize it for various problems in physical chemistry. The aim of the second part of this book is therefore to review the possible applications of gas chromatographic headspace analysis in this branch of chemistry. The major part of the discussion is concerned with the small number of publications which already exist on this topic, and the final section deals with other applications which appear to be feasible, but which have yet to be proved experimentally.

2.2. PROVEN APPLICATIONS

The experimental studies which have so far been published on the determination of physico-chemical constants by means of gas chromatographic headspace analysis all come within the sphere of the 'thermodynamics of phase equilibria',[1-5] and more specifically within the narrower category of 'vapour–liquid equilibria'.[6-8] The range of applications defined by the latter term is, of course, very limited, but nevertheless it is of considerable importance in so far as it provides the essential data on substances required to select the optimum conditions for the distillative separation processes of rectification, azeotropic rectification and extractive rectification, which represent the most frequently employed thermal separation processes in the chemical and petroleum industries. This topic has also attained some importance because information obtained from the measurements made can influence the fundamental theories of the intermolecular forces in liquids.

2.21. Vapour–Liquid Equilibrium Data for the Development of Separation Processes Based on Rectification

2.211. Introduction

The vapour which is in equilibrium with a mixture of liquids generally has a different composition from that of the liquid phase. The resulting difference in concentrations is utilized in separation processes in which rectification is used to resolve mixtures into their components. In the development of a process the equipment required can be optimized by means of calculations if vapour–liquid equilibrium data are known for the mixture being separated. However, it is not yet possible to calculate the equilibrium constants reliably in advance, since neither the present state of the theory nor the existing molecular data are adequate.[9] The vapour–liquid equilibrium data required therefore still have to be determined experimentally, if from the results of earlier measurements they are not available in the literature.[6,10-13]

A wide variety of equipment and measurement techniques has been developed for this purpose.[6] Depending on whether the two phases are agitated or are stationary during the period in which the equilibrium is established, the methods were classified into dynamic and static methods. The measurement techniques developed by analytical chemists using the Multifract F40 and F42[14] instruments are static methods, just like the procedures[15–17] which were suggested by some thermodynamicists using gas chromatographic headspace analysis.

2.212. Thermodynamic variables and parameters

The equilibrium behaviour of a binary mixture, i.e. one consisting of two components 1 and 2, is generally described in terms of the following measurable quantities: the temperature T, the total pressure P, the mole fractions x_1 and x_2 of the two components 1 and 2 in the liquid phase, and the mole fractions y_1 and y_2 of the same compounds in the vapour phase. It is conventional to use the subscript 1 for the lower-boiling, more volatile of the two components.

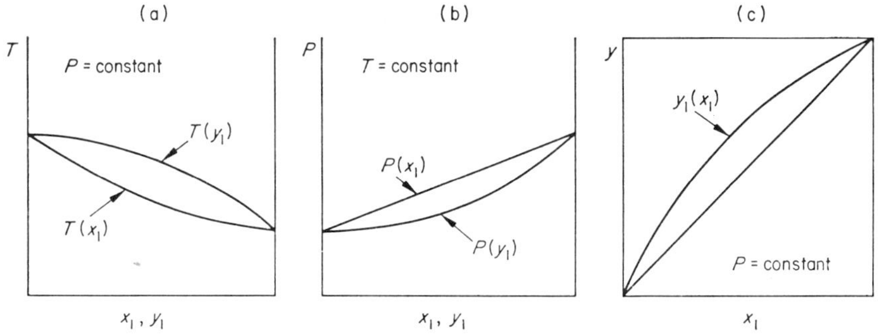

Fig. 58 Representation of vapour–liquid equilibria: (a) boiling point diagram; (b) vapour pressure diagram; (c) vapour–liquid equilibrium diagram.

The state of a binary mixture in which the vapour and liquid are in equilibrium can be represented by two surfaces having the spatial coordinates P, T, x_1 and P, T, y_1.[18] In a two-dimensional representation one of the variables must therefore be taken as a constant. It is customary to draw the vapour–liquid phase diagrams, in which the pressure or the temperature is constant (Figs. 58a and 58b). Another representation of the vapour–liquid equilibrium, which is frequently employed in relation to distillation columns, is known as the vapour–liquid equilibrium diagram. It shows the variation of the composition of the vapour at constant pressure with the composition of the liquid (Fig. 58c).

It is not necessary to measure all of the four quantities P, T, x_1 and y_1 in order to define the equilibrium position of the vapour–liquid mixture. The thermodynamics of phase equilibria, which will be discussed in more detail later, provides the possibility of calculating one of these quantities from thermodynamic relationships. This enables the equilibrium measurements to be simplified considerably, and this is particularly valuable in the case of static methods.

The *relative volatility* α, which is also called the *separation factor*, is a quantity which can frequently be used to estimate the separability of two substances 1 and 2 by distillation:

$$\alpha_{12} = \frac{y_1}{x_1}\frac{x_2}{y_2} \tag{11}$$

With increasing divergence of the relative volatility defined by Eq. (11) from unity the difference in composition between the vapour phase and the liquid phase increases, and as a consequence less effort is required to separate the components. The mole fraction y_1 in the vapour in equilibrium with the liquid may be calculated from the mole fraction in the liquid if the relative volatility is known:

$$y_1 = \frac{\alpha_{12}x_1}{1 + x_1(\alpha_{12} - 1)} \tag{12}$$

If the vapour and the liquid behave as ideal mixtures, then the relative volatility is equal to the ratio between the vapour pressures p_{01} and p_{02} of the pure substances, and is therefore independent of the composition of the mixture:

$$\alpha_{12} = (p_{01}/p_{02}) \tag{13a}$$

However, this 'ideal separation factor' is rarely found, and then almost only for mixtures of isomers or homologues.

On the other hand, the liquid phase frequently behaves as a real mixture, while the vapour exhibits ideal behaviour. In this case the difference between the relative volatility and the ideal separation factor may be described in terms of the *activity coefficients* γ_1 and γ_2:

$$\alpha_{12} = (p_{01}/p_{02})(\gamma_1/\gamma_2) \tag{13b}$$

The expression for the relative volatility is no longer so simple and comprehensive if the vapour phase must also be considered as a real mixture and the influence of the pressure on the liquid phase has to be taken into account. This can be done mathematically by introducing two correction factors ϕ_1 and ϕ_2:

$$\alpha_{12} = (p_{01}/p_{02})(\gamma_1/\gamma_2)(\phi_2/\phi_1) \tag{13c}$$

The following expressions are obtained for the correction factors ϕ_1 and ϕ_2:

$$\phi_1 = \exp\frac{(B_{11} - V'_{01})(P - p_{01})}{RT} \tag{14a}$$

$$\phi_2 = \exp\frac{(B_{22} - V'_{02})(P - p_{02})}{RT} \tag{14b}$$

if for the purpose of simplification it is assumed that (1) the vapours of the mixtures and of the pure compounds may be described sufficiently accurately by the virial equation of state expressed in terms of volumes, in which terms after that including the second virial coefficient B are neglected; (2) the molar volumes

V'_{01} and V'_{02} of the pure liquid compounds are incompressible throughout the pressure range from 0 to P; (3) the vapour mixture exhibits the behaviour of an ideal mixture of real gases (Lewis–Randall rule); (4) the activity coefficients γ_1 and γ_2 in Eq. (13c) are referred to the pure liquid compounds at the pressure and temperature of the mixture as their standard states.

In Eqs. (14a) and (14b) R is the molar gas constant and B_{11} and B_{22} are the second virial coefficients of the pure compounds.

The virial coefficients B_{11} and B_{22} can be estimated by a variety of methods starting from the critical data of the pure substances.[7,8] Redlich[19] has proposed the relationship

$$B = 0.0867\,\frac{RT_c}{P_c} - 0.4278\,\frac{RT_c}{P_c}\,(T_c/T)^{1.5} \tag{14c}$$

for the approximate calculation of the second virial coefficient B from the critical temperature T_c and the critical pressure P_c. The activity coefficient γ_i of a component i in the mixture is defined as being equal to the ratio of the activity a_i of the component to its mole fraction x_i:

$$\gamma_i = a_i/x_i \tag{15a}$$

The activity a_i is in turn defined as the ratio of the fugacity f_i of component i in the state in question to the fugacity f_{0i} of the same component i at the same temperature in a defined standard state:

$$a_i = f_i/f_{0i} \tag{15b}$$

For mixtures of liquids the pure liquid is chosen as an appropriate standard state.

The fugacity is a state function having the dimensions of pressure. If the saturated vapour of component i behaves as an ideal gas both in the pure state and in the mixture, then f_i is identical with the partial pressure p_i, and f_{0i} is equal to the vapour pressure p_{0i}. However, if it behaves as a real gas, these two pairs of values differ from each other. These differences can be expressed in terms of the corresponding *fugacity coefficients* φ_i and φ_{0i}:

$$\varphi_i = f_i/p_i \tag{16a}$$

$$\varphi_{0i} = f_{0i}/p_{0i} \tag{16b}$$

The fugacity coefficients are themselves expressed by exponential functions, and by use of the partial molar volumes V''_i and V''_{0i} of component i in its state in the mixture and in its pure state they can be written:

$$\varphi_i = \exp\left[\frac{1}{RT}\int_0^P \left(V''_i - \frac{RT}{P}\right) dP\right] \tag{17a}$$

$$\varphi_{0i} = \exp\left[\frac{1}{RT}\int_0^P \left(V''_{0i} - \frac{RT}{P}\right) dP\right] \tag{17b}$$

The activity coefficient itself, and also its variation with temperature and pressure, can be described in terms of the excess functions, which have proved to be extremely valuable in the thermodynamic treatment of phase equilibria:

$$\ln \gamma_i = (G_i - G_{0i})/RT = G_i^E/RT \tag{18}$$

$$(\delta \ln \gamma_i/\delta T)_{P,x} = -(H_i - H_{0i})/RT^2 = -H_i^E/RT^2 \tag{19}$$

$$(\delta \ln \gamma_i/\delta T)_{T,x} = (V_i - V_{0i})/RT = V_i^E/RT \tag{20}$$

where G_i, H_i and V_i are the partial molar values of the *free enthalpy*, the *enthalpy* and the *volume* of component i in its state in the mixture, G_{0i}, H_{0i} and V_{0i} are the corresponding molar quantities for component i in the standard state, and G_i^E, H_i^E and V_i^E are the corresponding partial molar excess functions of component i in its state in the mixture. The standard state here is the pure liquid component i at the same pressure and temperature as the mixture.

The variation of the activity coefficients γ_i with the composition of the mixture can be described by the empirical, or semi-empirical, equations which have been proposed by Wohl,[20] Orey and Prausnitz,[21] Renon and Prausnitz[22] and others.[23] These equations are solutions of the differential form of the Gibbs–Duhem equation:

$$\sum_i x_i \, d \ln \gamma_i = 0 \qquad [T, P] \tag{21}$$

which is written here in the form it takes at constant temperature and pressure. This applies to closed systems consisting of heterogeneous mixtures (e.g. vapour/liquid) and restricts the number of degrees of freedom in each phase by one. Moreover, it leads to a considerable reduction in the range of experimental measurements which are necessary to establish the properties of a mixture as a function of its composition. Its most important application is in the testing of experimental data for thermodynamic consistency, which is considered to be demonstrated if the data satisfy this equation.

A particularly useful formulation of the Gibbs–Duhem equation was derived by Van Ness[2] for the case of binary mixtures:

$$d\frac{\bar{G}^E}{RT} = -\frac{\bar{H}^E}{RT^2}\,dT + \frac{\bar{V}^E}{RT}\,dP + \ln\frac{\gamma_1}{\gamma_2}\,dx_1 \tag{22}$$

in which $\bar{G}^E$, $\bar{H}^E$ and $\bar{V}^E$ are the mean values of the molar excess functions of the free enthalpy, the enthalpy and the volume. They are related to the partial molar excess functions, which are given by Eqs. (18), (19) and (20), by the following expressions:

$$\bar{G}^E = \sum_{i=1}^{2} x_i G_i^E \tag{23a}$$

$$\bar{H}^E = \sum_{i=1}^{2} x_i H_i^E \tag{23b}$$

$$\bar{V}^E = \sum_{i=1}^{2} x_i V_i^E \tag{23c}$$

Equation (22) gives rise to two integration tests, one of which is for the case of isothermal, non-isobaric equilibrium measurements, and the other of which applies to isobaric, non-isothermal measurements. On integrating Eq. (22) over the whole range of mixtures from $x_1 = 0$ to $x_1 = 1$, the value of the left-hand side of the equation becomes equal to zero, because the mean molar excess free enthalpy $\bar{G}^{\mathrm{E}}$ vanishes at the limits $x_1 = 0$ and $x_1 = 1$, and this gives the two conditions:

$$\int_{x_1=0}^{x_1=1} \ln \frac{\gamma_1}{\gamma_2} \, \mathrm{d}x_1 = -\int_{x_1=0}^{x_1=1} \frac{\bar{V}^{\mathrm{E}}}{RT} \, \mathrm{d}P \quad \text{(isothermal case)} \tag{24a}$$

$$\int_{x_1=0}^{x_1=1} \ln \frac{\gamma_1}{\gamma_2} \, \mathrm{d}x_1 = +\int_{x_1=0}^{x_1=1} \frac{\bar{H}^{\mathrm{E}}}{RT^2} \, \mathrm{d}T \quad \text{(isobaric case)} \tag{24b}$$

The experimental data are considered to be thermodynamically consistent if these conditions are satisfied. In Eqs. (24a) and (24b) the standard states are again the pure liquids at the same pressure and temperature as the mixture. The ratio of the activity coefficients γ_1/γ_2 which is needed in Eqs. (24a) and (24b) is obtained from the experimental data by using Eq. (13b) or Eq. (13c). Thus, for example, in the isothermal case it can then be established whether the equation

$$\int_0^1 \ln \alpha_{12} \, \mathrm{d}x_1 - \int_0^1 \ln (p_{01}\phi_2/p_{02}\phi_1) \, \mathrm{d}x_1 = \int_{x_1=0}^{x_1=1} (\bar{V}^{\mathrm{E}}/RT) \, \mathrm{d}P \approx 0$$

is satisfied. The graphical procedure for performing this test of the consistency is shown in Fig. 59.

If all four of the quantities x_1, y_1, P (including p_{01} and p_{02}) and T have been measured directly under equilibrium conditions, the individual activity coefficients γ_1 and γ_2 for a binary system are obtained from the equations:

$$\gamma_1 = (y_1 P)/(x_1 p_{01}) \tag{25a}$$

$$\gamma_2 = (y_2 P)/(x_2 p_{02}) \tag{25b}$$

or, if correction factors for real gases have to be taken into account:

$$\gamma_1 = (y_1 P \phi_1)/(x_1 p_{01}) \tag{26a}$$

$$\gamma_2 = (y_2 P \phi_2)/(x_2 p_{02}) \tag{26b}$$

If for an isothermal system at equilibrium the pressure P has not been measured, the activity coefficients can usually no longer be determined so easily, but only in an indirect manner.

An indirect method of this kind[16] starting from Eq. (22) involves determination of $\bar{G}^{\mathrm{E}}$ for various values of x_1 within the range of compositions of the mixture by integration of the equation between the limits $x_1 = 0$ and x_1.

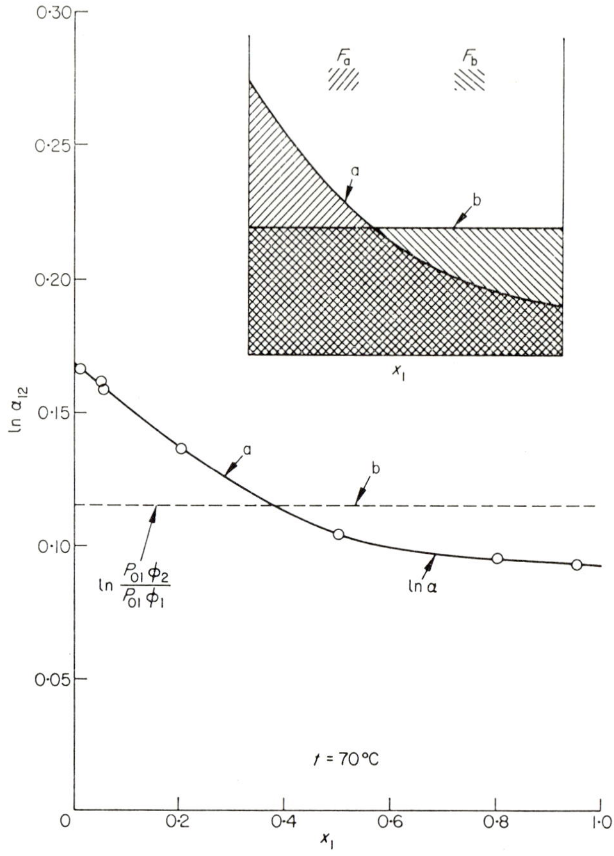

Fig. 59 Graphical test of isothermal experimental data for thermodynamic consistency.

Then use of Eqs. (27a) and (27b) finally leads to the expressions

$$\ln \gamma_1 = \frac{\bar{G}^E}{RT} + x_2 \ln \frac{\gamma_1}{\gamma_2} \tag{27a}$$

$$\ln \gamma_2 = \frac{\bar{G}^E}{RT} - x_1 \ln \frac{\gamma_1}{\gamma_2} \tag{27b}$$

The required ratio of the activity coefficients is again obtained from the experimental results by use of Eq. (13b) or Eq. (13c).

Another indirect method[17] starts from Eq. (13b) or Eq. (13c) by substituting in it the activity coefficient ratio γ_1/γ_2 derived from an empirical, or semi-empirical, expression representing the variation of the activity coefficient with the concentration, such as, for example, one of the equations due to Wohl,[20] Orey and Prausnitz,[21] Renon and Prausnitz[22] and others.[23] Then, by use of a numerical technique, the parameters in the equation employed are optimized so that the differences between the measured and calculated relative volatilities are at a minimum.

If the isothermal variation of the activity coefficients over the composition range of the mixture is known at different temperatures, then it follows that the isotherms of the *mean molar excess free enthalpy* $\bar{G}^E$ are also known, the individual points either having been determined from the integrated form of Eq. (22):

$$\bar{G}^E/RT = \int_{x_1=0}^{x_1} \left(\ln \alpha_{12} - \ln \frac{p_{01}\phi_2}{p_{02}\phi_1}\right) dx_1 \tag{22a}$$

neglecting the terms containing $\bar{V}^E$ (Fig. 60), or being given by the general equation:

$$\bar{G}^E/RT = \sum_{i=1}^{i=2} x_i \ln \gamma_i \quad [T, x] \tag{28}$$

for a given temperature and composition.

The *mean molar excess enthalpy* $\bar{H}^E$ is identical with the mean molar heat of mixing $\Delta\bar{H}$, and may be determined from the mean molar excess free enthalpy $\bar{G}^E$ by means of the relation

$$(d(\bar{G}^E/RT)/dT)_{x_1} = -\bar{H}^E/RT^2 + (\bar{V}^E/RT)(dP/dT)_{x_1} \tag{29}$$

which follows from Eq. (22) on differentiating with respect to the temperature T at constant mole fraction x_1. If the term in $\bar{V}^E$ is neglected, then the values obtained from

$$(d(\bar{G}^E/RT)/d(1/T))_{x_1} = \bar{H}^E/R \tag{29a}$$

are often accurate to within $\pm 15\%$.[24]

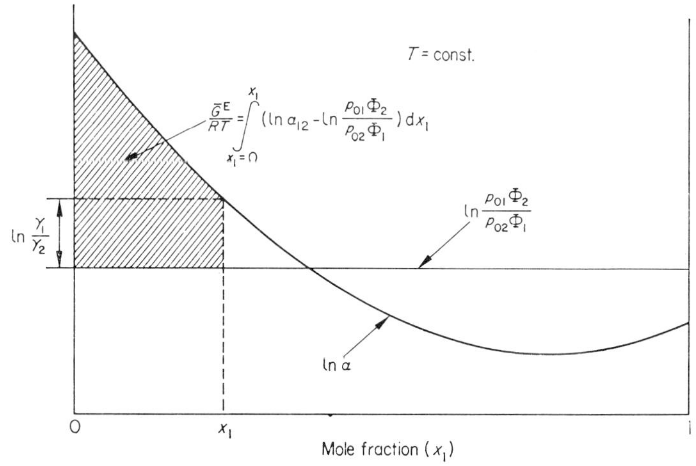

Fig. 60 Graphical determination of the mean molar excess free enthalpy $\bar{G}^E$.

Finally, by means of the Gibbs–Helmholtz equation:

$$\bar{H}^{E} = \bar{G}^{E} + T\bar{S}^{E} \tag{30}$$

the *mean molar excess entropy* $\bar{S}^{E}$ can also be determined.

If the activity coefficients for the binary mixture are known, the vapour pressure P, which will probably not have been measured, can be calculated from the equation

$$P = \gamma_1 x_1 p_{01} + \gamma_2 x_2 p_{02} \tag{31a}$$

or, if real gas correction factors have to be taken into account, from

$$P = \frac{\gamma_1 x_1 p_{01}}{\phi_1} + \frac{\gamma_2 x_2 p_{02}}{\phi_2} \tag{31b}$$

in which all the quantities are now known. The vapour pressure P in the exponential term in the correction factors ϕ_1 and ϕ_2 (cf. Eqs. (14a) and (14b)) does not have to be known very accurately, and it can be estimated approximately, for example, from Eq. (31a), in which for simplicity the activity coefficients γ_1 and γ_2 can each be taken equal to 1.

In addition, *isobaric boiling points* may be calculated by neglecting the variation of the activity coefficients with temperature, and varying the temperature at a given pressure and composition until Eq. (31a) or Eq. (31b) is satisfied. This involves using an empirical expression for the vapour pressures p_{01} and p_{02} of the pure liquids 1 and 2, e.g. the Antoine equation:

$$\ln p = A^* - \frac{B^*}{C^* + \vartheta} \tag{32}$$

in which A^*, B^* and C^* are constants characteristic of the substance, and ϑ is the temperature (°C).

The relative volatility (separation factor) of two substances 1 and 2 as defined by Eq. (11) can be affected by any additional substance (solvent) 3 which dilutes the liquid mixture of substances 1 and 2. The resulting effect, which is utilized in extractive and azeotropic distillation, is called the *improvement factor* β, and is defined as the ratio between the relative volatilities of substances 1 and 2 in the presence and absence of the additional solvent 3:[23,25,26]

$$\frac{(\alpha_{12})_{123}}{(\alpha_{12})_{12}} = \beta \tag{33}$$

In the limiting case of infinite dilution of substances 1 and 2 in 3 Eq. (33) takes the form

$$\frac{(\alpha_{12})_{12\infty 3}}{(\alpha_{12})_{12}} = \beta^* \tag{33a}$$

The effect of the additional solvent on the relative volatility may also be described in terms of the *selectivity* S, which is defined as the ratio between the activity coefficients of the substances 1 and 2 dissolved in the solvent 3:[27–31]

$$(\gamma_1/\gamma_2)_{123} = S \tag{34}$$

In the limiting case of infinite dilution of substances 1 and 2 in 3 Eq. (34) is used in the form

$$(\gamma_1/\gamma_2)_{12\infty 3} = S^* \tag{34a}$$

Application of Eq. (13b) to the ternary mixture 1–2–3 and to the binary mixture 1–2 in combination with Eqs. (33) and (33a) leads to the expressions

$$(\gamma_1/\gamma_2)_{123}/(\gamma_1/\gamma_2)_{12} = \beta \tag{35}$$

$$(\gamma_1/\gamma_2)_{12\infty 3}/(\gamma_1/\gamma_2)_{12} = \beta^* \tag{35a}$$

Substitution of Eqs. (34) and (34a) into Eqs. (35) and (35a) gives the following relationships between the improvement factor and the selectivity:

$$S = \beta(\gamma_1/\gamma_2)_{12} \tag{36}$$

$$S^* = \beta^*(\gamma_1/\gamma_2)_{12} \tag{36a}$$

Since relative volatilities and activity coefficients depend primarily on the temperature and on concentrations (the variation with pressure is more or less negligible in the low pressure region), the temperature and the ratios of the concentrations of substances 1 and 2 must be kept constant to satisfy Eqs. (33)–(36a). The magnitude of the effect on the relative volatility of two substances 1 and 2 caused by the addition of another substance 3 may be derived by calculation, estimation or measurement.

The calculations are based on experimental data on the vapour–liquid equilibria of the binary sub-systems, 1–2, 2–3 and 1–3 in the ternary system 1–2–3, and use the equations proposed by Wohl,[20] Orey and Prausnitz,[21] Renon and Prausnitz[22] and others[23] to derive the selectivity S and the improvement factor β.

The estimations, which start from the physical data on the pure substances, are based on the equations proposed by Pierotti, Deal and Derr,[32] Hildebrand and Scott,[33] Weimer and Prausnitz,[34] Helpinstill and Van Winkle[35] or Null and Palmer,[36] and give the selectivity at infinite dilution S^*.

The measurements are usually performed with vapour–liquid equilibrium apparatus,[6] or gas chromatographically with columns in which the additional substance 3 serves as the stationary phase.[37–44] Whereas the improvement factor β and the selectivity S can be obtained from measurements of the vapour–liquid equilibria, the gas chromatographic measurements provide only the selectivity at infinite dilution S^*.

Since systems exhibiting normal[45] and abnormal[46,47] variations of the selectivity S with the concentration of the solvent exist, the values obtained by the chromatographic method described (solvent as the stationary phase), and by estimation (from the physical data on the pure substances), for the selectivity at infinite dilution S^* can not be used to calculate the selectivity at the finite dilutions which are used in extractive distillations. Reliable values for the improvement factor β and the selectivity S at solvent concentrations which are actually used in extractive distillations have therefore only been obtained up to now by measurements of vapour–liquid equilibria, but these are seldom available from the literature and measurements must usually be carried out when data are required.

2.213. Determination of thermodynamic quantities

The original suggestion for measuring vapour–liquid equilibrium data by a method which would now be described as gas chromatographic headspace analysis was made by Wichterle and Hala.[15] These authors emphasized the special advantages of this method in that it requires a liquid sample of only 2 cm^3, it is not necessary to wait for a static equilibrium to be completely established, and a considerable amount of quantitative data can be obtained without causing significant changes in the concentrations in the liquid phase. The simple and ingenious technique makes the method particularly suitable for measurements on isothermal vapour–liquid equilibria in multicomponent systems, and for the systematic selection of solvents for extractive distillation.

The equilibration vessel is a conical glass flask (Fig. 61) having a glass temperature-controlled jacket through which liquid is circulated. It is fitted with a stopcock which provides the means of opening and closing the flask, and also enables samples to be taken from the gas phase. The dual function of the

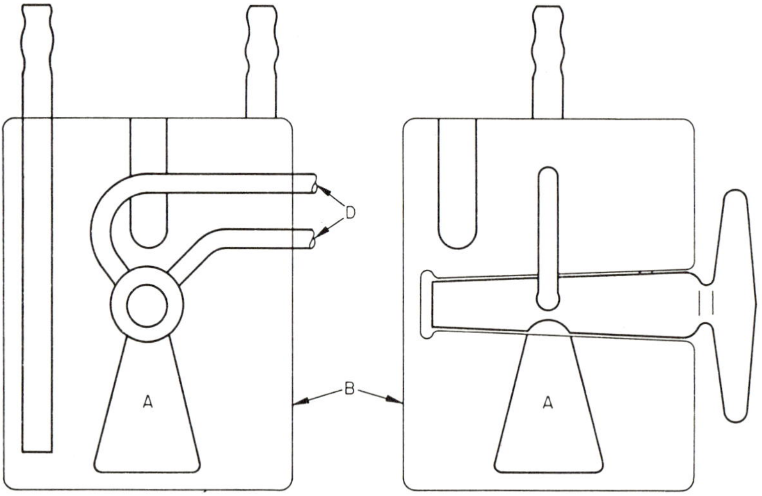

Fig. 61 Equilibration vessel used by Wichterle and Hala: A, equilibration vessel; B, temperature-controlled jacket; C, stopcock key; D, carrier gas inlet and outlet tubes.

stopcock is provided by a small groove cut in to the centre of its key, so that, depending on the position of the key, the groove is connected either to the closed vessel or to the column of an analytical gas chromatograph with the other side of the key isolating the flask (Fig. 62). The stopcock key must be removed to introduce or remove liquid samples from the flask.

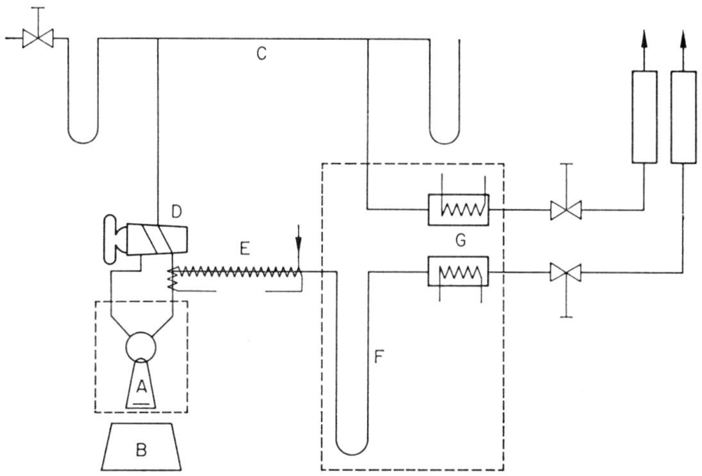

Fig. 62 Wichterle and Hala's apparatus: A, equilibration vessel (thermostatted); B, magnetic stirrer; C, carrier gas line; D, two-way stopcock; E, electric heater; F, chromatographic column; G, chromatographic detector (catharometer).

While equilibrium is being established, the groove, which has a volume of 0.5 cm^3, is connected with the flask, which has a volume of about 10 cm^3. The liquid phase is agitated with an electromagnetic stirrer, contact between the liquid and the stopcock being avoided. The temperature of the stopcock and the equilibration vessel is held constant to ± 0.05°C. The stopcock is carefully polished, but is not greased. The tube connecting the equilibration vessel to the chromatograph is electrically heated to prevent condensation occurring. The temperature of the chromatograph is held constant to ± 0.1°C at about 1°C above that of the equilibration vessel. A sample of the vapour phase is taken by rotating the stopcock key through 180° so that the carrier gas transports the vapour from the groove in to the chromatograph. A sample of the liquid phase is also analysed by gas chromatography and the ratios between the peak areas are compared. Since the liquid phase is obtained by mixing weighed amounts of the pure substances, it has a known composition. The relative volatility can therefore be calculated by comparing the ratios between the peak areas of the components in the liquid and vapour phases.

The equilibration apparatus has been tested with the *n*-heptane/toluene system at 90°C. To provide a comparison the isobaric data in the literature were first converted into isothermal data. Values of the ratio between the activity coefficients γ_1/γ_2 were then calculated, and by use of Eqs. (13b) and (12) the relative volatilities and, hence, the mole fractions in the vapour phase were calculated for each composition of the mixture at which experimental measurements were made. Comparison of the data showed good agreement between the experimental and the published values.

The method these authors proposed is experimentally very simple and yields reliable results in a relatively short time.

Dietrich[16] has recently employed gas chromatographic headspace analysis to investigate the vapour–liquid equilibrium in the binary system benzene/1,2-dichloroethane at various temperatures, and has shown that very accurate and thermodynamically consistent experimental results can be obtained by a static method if appropriate consideration has been given to the design of the apparatus and its ancillary equipment. In contrast to the conventional techniques, which at low pressures give only approximate values from measurements on vapour–liquid equilibria, the apparatus (Fig. 63) developed by Dietrich can be used satisfactorily for measurements at these pressures. In this method the actual pressure is not measured, but is calculated as previously done in the procedure used by Wichterle and Hala. This has the advantage that it is not necessary to degas the liquid mixture and therefore its composition can be defined very simply by weighing out the two pure substances. The liquid should be continuously stirred while the equilibrium is being established.

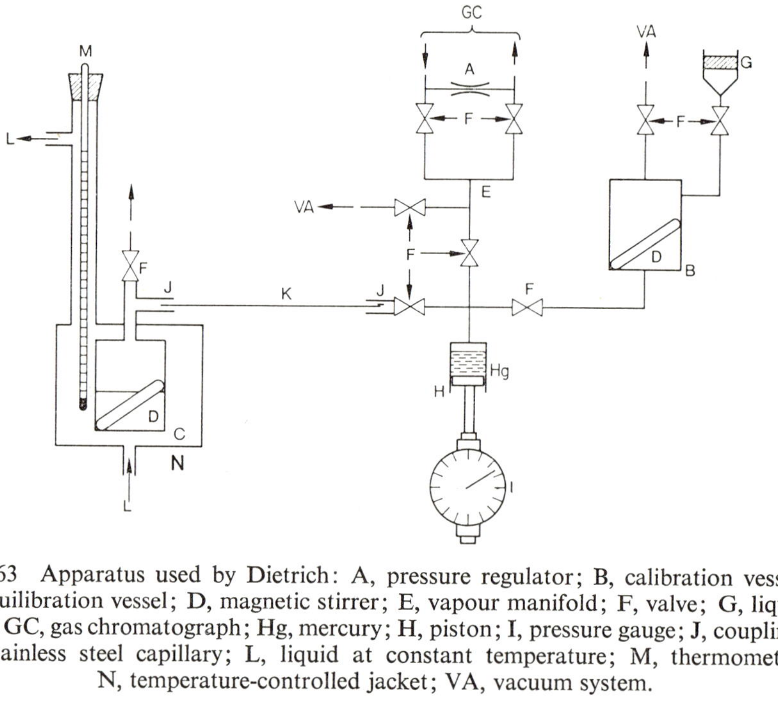

Fig. 63 Apparatus used by Dietrich: A, pressure regulator; B, calibration vessel; C, equilibration vessel; D, magnetic stirrer; E, vapour manifold; F, valve; G, liquid inlet; GC, gas chromatograph; Hg, mercury; H, piston; I, pressure gauge; J, coupling; K, stainless steel capillary; L, liquid at constant temperature; M, thermometer; N, temperature-controlled jacket; VA, vacuum system.

The vapour pressures of the pure substances are measured directly in the same apparatus, but the system and the liquid must have been previously degassed. The method of performing these vapour pressure measurements is based on the same principle as the well-known static method of measuring vapour pressures with the isoteniscope.[6]

The equilibration vessel was a 50 mm long cylinder with an internal diameter of 15 mm, enclosed in a glass temperature-controlled jacket through which liquid was circulated. The temperature of the liquid was controlled by a modified Haake thermostat, the long term drifts in the temperature being $\pm 5 \times 10^{-3}$°C. The equilibration vessel contained a glass-covered stirrer magnet, 35 mm long, which could be rotated in a vertical plane at different speeds by a magnetic stirrer located beyond the temperature-controlled jacket.

After equilibrium had been established between the phases, a series of vapour samples was removed from the headspace of the equilibration vessel, by causing a drop in the pressure in a connecting tube between the vessel and a sample chamber, the volume of which could be changed by moving a piston covered with a layer of mercury. By operating several valves each of the samples which had been introduced into the sample chamber was transferred into a vapour-lock from which it was transported into the gas chromatograph (Perkin-Elmer, Bodenseewerk, model F 6/2) more or less as a slug of vapour in the carrier gas. Analysis of the vapour sample gave the peak areas F''_1 and F''_2 of the two components in the mixture.

The sample chamber was connected via a tube incorporating a valve to a second vessel, which was used to calibrate the apparatus. A small quantity of liquid in which the mole fraction x_1 was known was poured into this vessel and completely vaporized. A series of vapour samples was then removed from the vessel in the same way as they were taken from the equilibration vessel, and analysed in the gas chromatograph. The analysis of the completely vaporized liquid sample gave the peak areas F'_1 and F'_2 of the two components.

The quantities calculated by Dietrich from his experimental results included the following: the relative volatility α_{12}, the mean molar excess free enthalpy $\bar{G}^E$, the activity coefficients γ_1 and γ_2, the total pressure P, the isobaric boiling point T, the mean molar excess enthalpy $\bar{H}^E$ and the mean molar excess entropy $\bar{S}^E$.

To determine the relative volatility α_{12} in the benzene/1,2-dichloroethane system Dietrich made use of the relationship

$$\ln \alpha_{12} = \ln (F''_1/F''_2) - \ln (F'_1/F'_2) + \ln (A'/A'') \qquad (37)$$

A is termed the calibration factor, and is the ratio c_1/c_2 between the proportionality constants $c_1 = F_1/n_1$ and $c_2 = F_2/n_2$, n_1 and n_2 being the numbers of moles. As a consequence of the incomplete resolution of the substances in the gas chromatographic column employed, A was a function of both the composition and of the size of the sample. A' and A'' are the calibration factors corresponding to the liquid sample and the co-existing vapour sample. The experimental values substituted in the equation, the peak areas and the calibration factors, always refer to the same sample size, and this can be achieved by interpolation of measurements made with different sized samples.

Figure 64 shows the experimental values of the relative volatility α_{12} at various temperatures for the benzene/1,2-dichloroethane system as a function of the mole fraction x_1 in the liquid.

Using the value calculated for the relative volatility, Eq. (12) was used to determine the mole fractions y_1 and x_1 in the co-existing phases, and they are presented graphically as the difference $y_1 - x_1$ in Fig. 65.

Equation (13c) was chosen for the mathematical description of the relative volatility, the correction factors ϕ_1 and ϕ_2 being given by Eqs. (14a) and (14b),

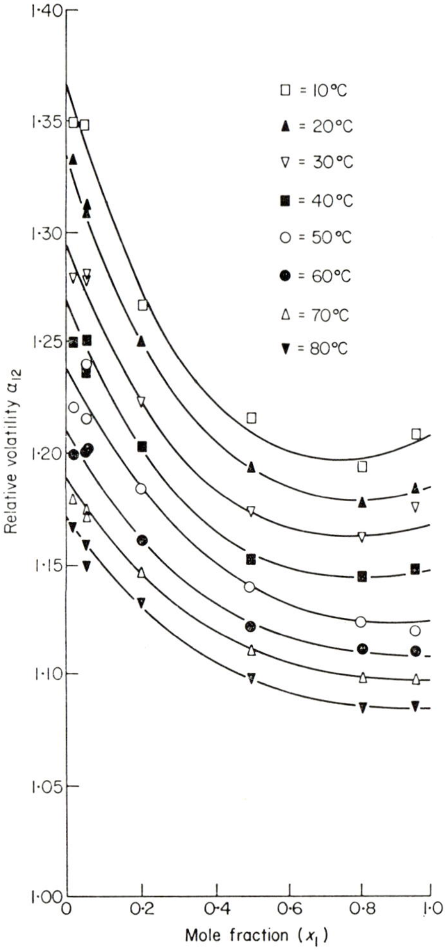

Fig. 64 The relative volatility α_{12} in the benzene/1,2-dichloroethane system measured by Dietrich.

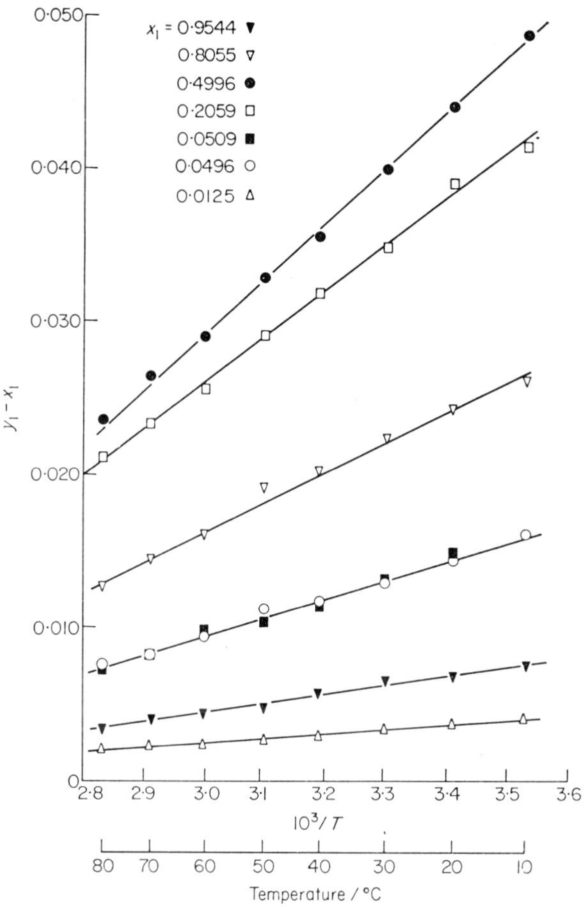

Fig. 65 The variation with temperature of the difference $y_1 - x_1$ between the concentrations for the vapour–liquid equilibrium in the benzene/1,2-dichloroethane system (Dietrich).

respectively. The virial coefficients required were calculated approximately from Eq. (14c), and the pressure P was estimated by means of Raoult's and Dalton's laws. The experimental results obtained under isothermal conditions were tested to see whether they satisfied Eq. (24a), in which the term including $\bar{V}^E$ is negligible, and in which an expression for γ_1/γ_2 derived from Eq. (13c) was substituted. The procedure is illustrated in Fig. 59, and showed that the results were thermodynamically consistent.

Since the vapour pressure was not measured with sufficient accuracy, the activity coefficients γ_1 and γ_2 were determined from Eqs. (27a) and (27b), the mean molar excess free enthalpy $\bar{G}^E$ having been first obtained from Eq. (22a). Equation (29a) was then used to obtain the mean molar excess enthalpy $\bar{H}^E$

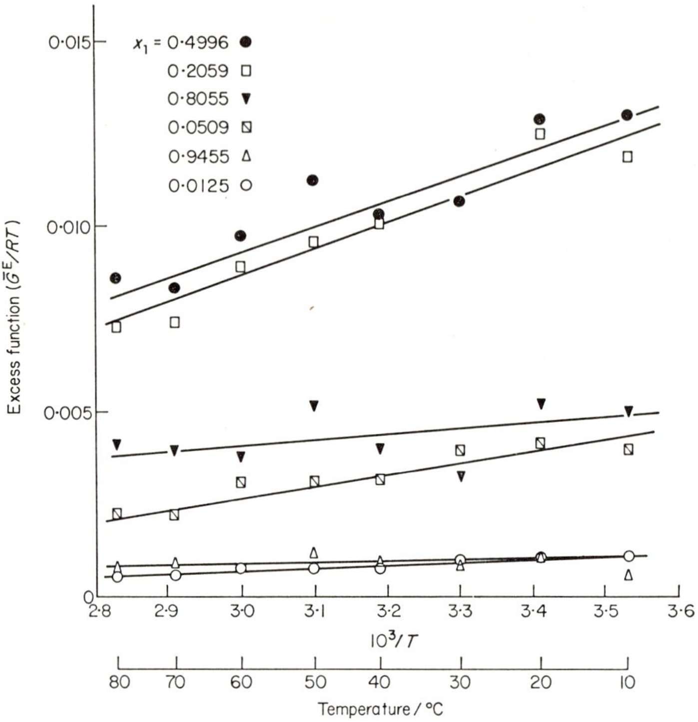

Fig. 66 The variation of the excess function $\bar{G}^E/RT$ with the temperature in the benzene/1,2-dichloroethane system (Dietrich).

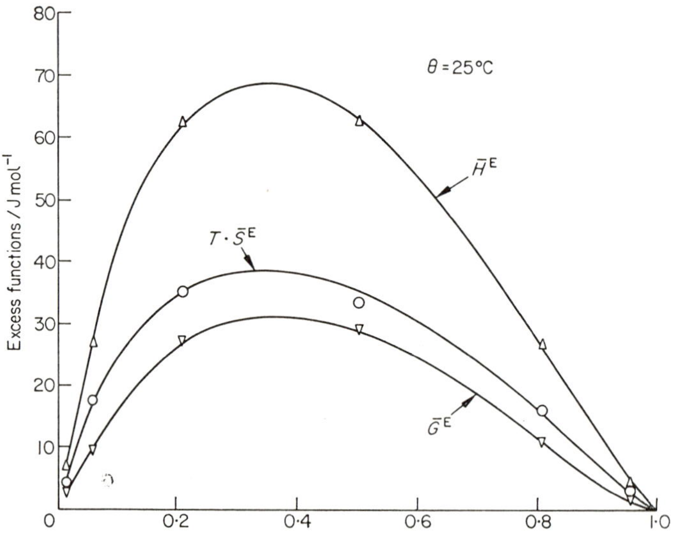

Fig. 67 The excess functions $\bar{G}^E$, $\bar{H}^E$ and $T\bar{S}^E$ measured for the benzene/1,2-dichloroethane system at 25°C by Dietrich.

graphically from the slope of the function $(\bar{G}^{E}/RT)_{x_1}$ which is plotted in Fig. 66. Finally, the mean molar excess entropy $\bar{S}^{E}$ was obtained from Eq. (30). The results of these calculations are plotted in Fig. 67.

The values obtained for the activity coefficients γ_1 and γ_2 were used in conjunction with Eq. (31b) to calculate the vapour pressure P, which was not measured with the other data for the isothermal equilibrium. Isobaric boiling points were also calculated. The results were compared with those of other authors, and it was shown that these latest values are among the most accurate available.

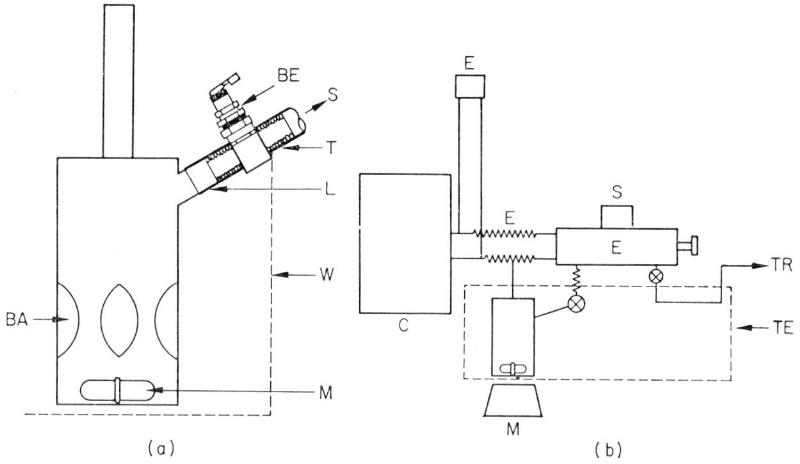

Fig. 68 The apparatus used by Montfort. (a) Equilibration vessel: BA, baffles; BE, bellows valve; L, latex connector; M, magnetic stirrer; S, to sampling valve; T, Teflon connector; W, water-bath. (b) Diagram of whole apparatus: C, chromatograph; E, electric heaters; M, magnetic stirrer; S, sampling valve; TE, thermostatted bath; TR, trap and vacuum pump.

Finally, it should be noted that the gas chromatographic headspace analysis procedure demonstrated by Dietrich certainly provides very accurate values for all of the quantities which can be obtained from measurements on vapour–liquid systems at equilibrium, but that a considerable amount of effort is required to obtain the most important quantities, i.e. the individual values of γ_1 and γ_2.

Another static method (Fig. 68) of measuring vapour–liquid equilibrium data by means of gas chromatographic headspace analysis has been developed by Montfort.[17] This is similar to the technique described by Dietrich in that only the relative volatility is measured by gas chromatographic analysis of the two phases in equilibrium. Optimum values for the constants in the Wilson equation can be obtained, without using a real gas correction factor, by numerical fitting of the calculated and experimentally determined relative volatilities. The Wilson equation then enables all of the thermodynamically relevant quantities to be calculated for the mixture under the conditions of vapour–liquid equilibrium.

For 80 equilibrium positions at which measurements were made in the six binary systems benzene/toluene, toluene/ethanol, ethanol/*n*-heptane, toluene/acetonitrile, benzene/acetonitrile and benzene/*n*-heptane, the mean differences between the calculated and measured vapour concentrations and vapour pressures amounted to 0.8 mole % and 2.8 mm Hg, respectively.

In order to describe the two ternary systems acetonitrile/benzene/*n*-heptane and acetonitrile/benzene/toluene, Wilson parameters for the subsystem acetonitrile/*n*-heptane have been fitted by use of literature data on activity coefficients corresponding to the partially miscible mixture, while for the subsystem benzene/toluene the liquid mixture has been supposed ideal at the temperature considered. Several equilibrium points in the ternary systems were calculated by use of the completed set of Wilson parameters for the five binary subsystems. Reference measurements at these points demonstrated that the prediction is good: the average deviation of the experimental vapour concentrations was respectively 0.013 and 0.008 mole fraction for acetonitrile and benzene; the average deviation of the pressure was 12.4 mm Hg.

A further procedure for the static measurement of vapour–liquid equilibrium data by the gas chromatographic headspace analysis method was demonstrated by Kolb.[48] He carried out the measurements in a commercially manufactured gas chromatographic headspace analyser (model F42, Perkin-Elmer, Bodenseewerk, Überlingen), which differs significantly from the apparatus previously discussed in that it is equipped with an electropneumatic sampling system which ensures that a reproducible sample is removed from the headspace of the specimen vessel. For a given carrier gas inlet pressure and sampling time the carrier gas transfers a reproducible volume V of vapour from the vessel in which the vapour–liquid equilibrium has been established, into the gas chromatographic column. If the measurements are made in the linear range of the detector unit, and this must be confirmed by tests with, for example, a diluent gas, and the concentration of the vapour in the carrier gas is always kept low, then the peak area F_i obtained is proportional to the number of moles n_i of the component i. Therefore

$$F_i = c_i n_i \tag{38}$$

where the proportionality constant c_i depends on the detector employed and is specific for a given substance. It is possible to deduce the activity a_i of component i in the liquid sample from the gas chromatographic headspace analysis with the F40 and F42 instruments if an exact expression relating the number of moles n_i of component i to its fugacity f_i is available, which contains only quantities which can be measured or estimated. A relationship of this kind does exist; it was derived by Beattie starting from the Maxwell equations for the effect of volume on the internal energy and on the entropy,[49–51] and is best known in the form:[8]

$$RT \ln f_i = \int_V^{\infty} \left[\left(\frac{\delta P}{\delta n_i} \right)_{T,V,n_j} - \frac{RT}{V} \right] \mathrm{d}V - RT \ln \frac{V}{n_i RT} \tag{39}$$

If the virial equation describing the compressibility factor z as a power series of the reciprocal molar volume $\bar{V}$

$$z = \frac{P\bar{V}}{RT} = 1 + B/\bar{V} + C/\bar{V}^2 + \cdots \tag{40}$$

is used as the equation of state for real gases and gaseous mixtures with the progression being neglected after the second term, and the second virial coefficients being represented by the general expression

$$B = \sum_{i=1}^{m}\sum_{j=1}^{m} y_i y_j B_{ij} \tag{41}$$

then, from Eq. (39) the following relationship is obtained for the intermediate range of pressures:

$$n_i = (V/RT)\left\{f_i \exp\left[-\frac{2}{V}\sum_{j=1}^{m} n_j B_{ij}\right]\right\} \tag{42}$$

In principle the second virial coefficients B_{ij} in this equation either are known or can be estimated.[8,75]

A relationship between the peak area F_i and the fugacity f_i may be derived by substituting Eq. (42) into Eq. (38):

$$F_i = c_i(V/RT)\left\{f_i \exp\left[-\frac{2}{V}\sum_{j=1}^{m} n_j B_{ij}\right]\right\} \tag{43}$$

If in a series of experiments in which the F40 or F42 headspace analysis instrument is used at a given temperature T, equal volumes V of sample vapour at the same pressure (carrier gas inlet pressure) are removed from a headspace of constant volume, and analysed, the peak areas from different samples may be related to each other. Comparison of the value of F_i

$$F_i = c_i(V/RT)\left\{f_i \exp\left[-\frac{2}{V}\sum_{j=1}^{m} n_j B_{ij}\right]\right\}_{x_i \neq 1} \tag{43a}$$

obtained for a liquid mixture of component i ($x_i \neq 1$) with the value of F_{0i}

$$F_{0i} = c_i(V/RT)\left\{f_i \exp\left[-\frac{2}{V}\sum_{j=1}^{m} n_j B_{ij}\right]\right\}_{x_i = 1} \tag{43b}$$

obtained by using the pure liquid component i of the mixture as the sample, gives the ratio F_i/F_{0i}:

$$F_i/F_{0i} = \left\{f_i \exp\left[-\frac{2}{V}\sum_{j=1}^{m} n_j B_{ij}\right]\right\}_{x_i \neq 1} \Big/ \left\{f_i \exp\left[-\frac{2}{V}\sum_{j=1}^{m} n_j B_{ij}\right]\right\}_{x_i = 1} \tag{44}$$

In this equation f_i represents the fugacity of component i in the vapour phase. However, since at the time at which the sample is taken the vapour–liquid system is in equilibrium, then f_i also represents the fugacity in the liquid phase. Substitution of f_i/f_{0i} and transformation of Eq. (44) gives:

$$F_i/F_{0i} = (f_i/f_{0i}) \exp\left\{-\left[\frac{2}{V}\sum_{j=1}^{m} n_j B_{ij}\right]_{x_i \neq 1} + \left[\frac{2}{V}\sum_{j=1}^{m} n_j B_{ij}\right]_{x_i = 1}\right\} \tag{45}$$

On taking into consideration the fact that the activity a_i of component i in the liquid sample is by definition (Eq. (15b)) equal to the fugacity ratio f_i/f_{0i}, the following equation is obtained for the activity a_i referred to the pure liquid i as the standard state:

$$a_i = (F_i/F_{0i}) \exp\left\{+\left[\frac{2}{V}\sum_{j=1}^{m} n_j B_{ij}\right]_{x_i \neq 1} - \left[\frac{2}{V}\sum_{j=1}^{m} n_j B_{ij}\right]_{x_i = 1}\right\} \tag{46}$$

If the pure liquid i is chosen as the standard state, then the corresponding expression for the activity coefficient γ_i is

$$\gamma_i = (1/x_i)(F_i/F_{0i}) \exp\left\{+\left[\frac{2}{V}\sum_{j=1}^{m} n_j B_{ij}\right]_{x_i \neq 1} - \left[\frac{2}{V}\sum_{j=1}^{m} n_j B_{ij}\right]_{x_i = 1}\right\} \tag{47}$$

Since all of the virial coefficients higher than the second have been neglected, Eqs. (42)–(47) only hold at moderate densities, which in each case is less than half of the critical density.[7] Apart from this, the equations mentioned are not subject to any other limitations. If the temperature chosen for the measurements is sufficiently low, then of the m components in the vapour phase, nitrogen, oxygen and the carrier gas represent the major constituents, and their molar quantities n_j in the sample volume V will only vary slightly in a series of measurements. The real gas effect caused by the interaction between molecules of i and those of the inert gases should therefore be of reduced importance owing to partial compensation, so that the exponential function assumes a value which is not significantly different from unity:

$$a_i = F_i/F_{0i} \tag{46a}$$

$$\gamma_i = (1/x_i)(F_i/F_{0i}) \tag{47a}$$

The relationship is not as simple if specific interactions exist between the components of the mixture in the vapour phase, as, for example, in systems containing carboxylic acids.[52]

Kolb[48] has carried out measurements on the ethanol/n-heptane and chloroform/acetone systems. The peak areas obtained were normalized by means of a molar response factor so that the ratio of the normalized peak area to the partial pressure became the same for each of the components in the two systems. The response factors were determined by conventional gas chromatographic analysis of aliquots of standard mixtures of the liquids. Plots of the normalized peak areas against the respective mole fractions derived from the weighed amounts of the components in the corresponding mixtures gave peak area diagrams which were exactly the same as the familiar vapour pressure diagrams, except for the scale on the ordinate which covered a range different from unity.

The activity coefficients calculated by Kolb from his experimental measurements agree with the corresponding data in the literature,[72] and are shown graphically in Fig. 69. The equation used to calculate the activity coefficients

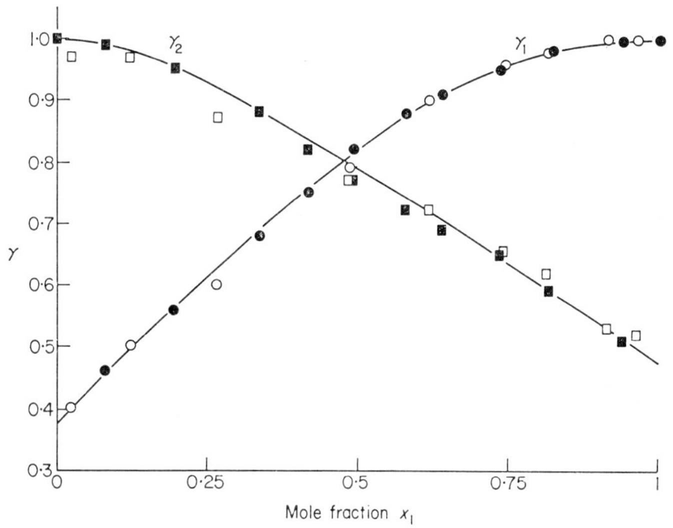

Fig. 69 Comparison of the activity coefficients measured by Kolb for the acetone/chloroform system (○, □) with data from the literature (●, ■)[72] at 36.5°C.

only differed from Eq. (47a) in that normalized peak areas were used instead of the measured peak areas. However, normalization of the peak areas is not necessary for the calculation of the activity coefficients, since if the measurements are made in the region of linear response of the gas chromatograph, the response factors are independent of the concentrations and therefore do not come into the calculation of the ratio of F_i and F_{0i}.

Rigorous application of the approximate equation (46a) to the experimental results which Hachenberg[53] obtained under isothermal conditions in similar gas chromatographic headspace analyses with the Multifract F40 instrument (Perkin-Elmer, Bodenseewerk) of the vinyl acetate/ethyl acetate system, gives the results shown in Fig. 70. The results presented in Table 12 were obtained for the same system by Schmidt using a classical method of making measurements on a vapour–liquid system at equilibrium. The results of Schmidt and Hachenberg are in agreement and demonstrate that the system exhibits ideal behaviour.[54]

TABLE 12

Vapour–liquid equilibrium data for the vinyl acetate/ethyl acetate system measured at 760 mm Hg (1.013 bar) and calculated assuming ideal behaviour. (Schmidt[55])*

	Measured		*Calculated (ideal behaviour)*	
x_1	y_1	ϑ(°C)	y_1	ϑ(°C)
0.190	0.205	76.25	0.213	76.22
0.390	0.420	75.35	0.425	75.32
0.790	0.830	73.60	0.813	73.60

* *The measurements were performed with an apparatus described by Röck and Sieg.*[56] *Equations* (12) *and* (13a) *were used in the calculations, which were based on the assumption that the liquid phase behaved as an ideal mixture. The vapour pressures of the pure substances were calculated from the following Antoine constants taken from the literature*[57]*: for vinyl acetate* (*subscript* 1), $A^* = 7.21010$, $B^* = 1296.130$, $C^* = 226.655$; *for ethyl acetate,* $A^* = 7.10179$, $B^* = 1244.951$, $C^* = 217.881$.

The experimental results which have been obtained by Kolb and Hachenberg in comparative gas chromatographic headspace analyses using the Multifract Automat (F40 and F42, Perkin-Elmer, Bodenseewerk) at pressures (carrier gas inlet pressures) between 1 and 2 bar, demonstrate that very accurate values for activities and individual activity coefficients can be rapidly determined with this instrument, using the very simple approximate Eqs. (46a) and (47a). The disadvantage that the equilibrium pressure can not be measured exactly in this

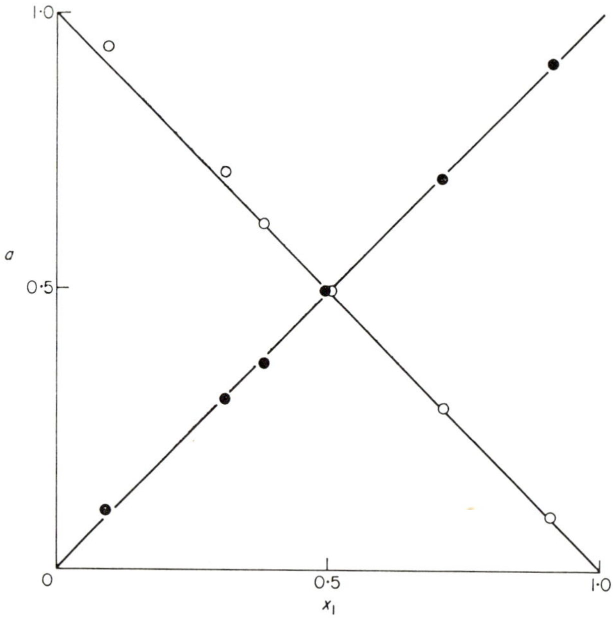

Fig. 70 The activity *a* in the vinyl acetate/ethyl acetate system at 50°C (unpublished results of H. Hachenberg).

static method of studying vapour–liquid equilibria is compensated for by the reproducibility of the sampling from the vapour phase. The method can be employed to particular advantage in measurements on multicomponent systems, for which the conventional methods are not even as accurate, but involves more effort.

Hachenberg and Schmidt[58] have shown that measurements made with a gas chromatographic headspace analyser (model F40 or F42, Perkin-Elmer, Bodenseewerk) form an excellent basis for the selection of solvents for extractive distillation. This technique provides a very rapid, simple and very accurate determination of the improvement factor β or β^* using the equations

$$\beta = \frac{(F_1/F_2)_{123}}{(F_1/F_2)_{12}} \tag{48}$$

$$\beta^* = \frac{(F_1/F_2)_{12\infty 3}}{(F_1/F_2)_{12}} \tag{48a}$$

in which F_1 and F_2 are the peak areas for compounds 1 and 2, which for comparison are measured in the headspace above liquid mixtures of compounds 1 and 2 in the presence and absence of compound 3 (Fig. 71). For the comparative measurements the pressure, temperature and ratios of the concentrations of the compounds 1 and 2 in the liquid mixtures must be kept the same. This approach has the great advantage that it is not necessary to determine the response factors or the concentrations.

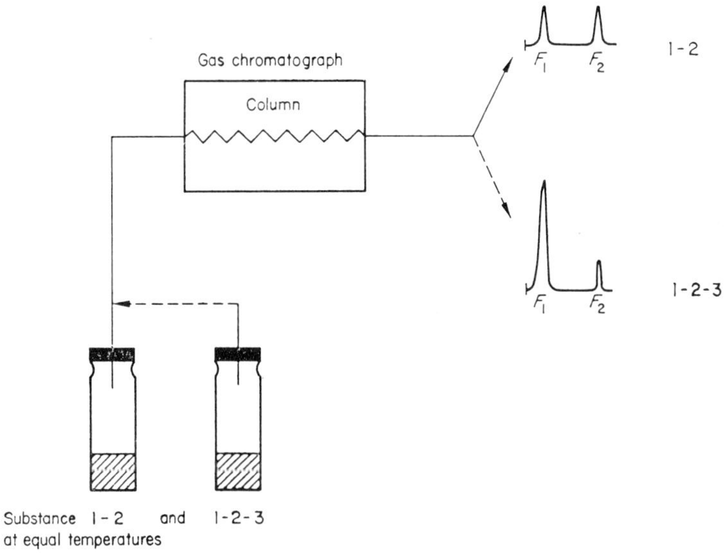

Fig. 71 Schematic representation of comparative gas chromatographic headspace analyses for the determination of the improvement factor β or β^*.

The preceding equations are derived by applying Eq. (11) to the two mixtures 1–2 and 1–2–3:

$$(\alpha_{12})_{123} = (y_1/y_2)_{123}(x_2/x_1)_{123} \tag{49}$$

$$(\alpha_{12})_{12\infty 3} = (y_1/y_2)_{12\infty 3}(x_2/x_1)_{12\infty 3} \tag{49a}$$

$$(\alpha_{12})_{12} = (y_1/y_2)_{12}(x_2/x_1)_{12} \tag{50}$$

and then substituting the relative volatilities obtained into Eqs. (33) and (33a):

$$\beta = \frac{(y_1/y_2)_{123}(x_2/x_1)_{123}}{(y_1/y_2)_{12}(x_2/x_1)_{12}} \tag{51}$$

$$\beta^* = \frac{(y_1/y_2)_{12\infty 3}(x_2/x_1)_{12\infty 3}}{(y_1/y_2)_{12}(x_2/x_1)_{12}} \tag{51a}$$

The corresponding ratios between the number of moles n_1/n_2 are then substituted for the ratios of the mole fractions in the gas phase y_1/y_2, and taking into account the equality of the ratios x_2/x_1 between the concentrations of compounds 1 and 2 in the liquid phases of the mixtures 1–2 and 1–2–3 gives:

$$\beta = \frac{(n_1/n_2)_{123}}{(n_1/n_2)_{12}} \tag{52}$$

$$\beta^* = \frac{(n_1/n_2)_{12\infty 3}}{(n_1/n_2)_{12}} \tag{52a}$$

Equation (38) is then used to replace the ratios of the number of moles of the compounds by the ratios of the peak areas F_1/F_2:

$$\beta = \frac{(F_1/F_2)_{123}(c_2/c_1)}{(F_1/F_2)_{12}(c_2/c_1)} \tag{53}$$

$$\beta^* = \frac{(F_1/F_2)_{12\infty 3}(c_2/c_1)}{(F_1/F_2)_{12}(c_2/c_1)} \tag{53a}$$

and finally the constants c_1 and c_2 are eliminated.

The change in the ratio of the peak areas which occurs on going from the two-component system 1–2 to the three-component system 1–2–3 can exhibit one of the three alternative forms shown in Fig. 72. These correspond to the fact that the improvement factors β and β^* derived from the ratios of the measured peak areas can either be equal to 1 or be greater or smaller than 1, depending on whether the solvent has no effect or has a positive or negative influence on the relative volatility of the two compounds 1 and 2. For extractive distillation it appears that those solvents which cause opposite effects (positive or negative) are equally suitable as extractive agents, so that in this case extractive distillation can be carried out in alternative ways.[59]

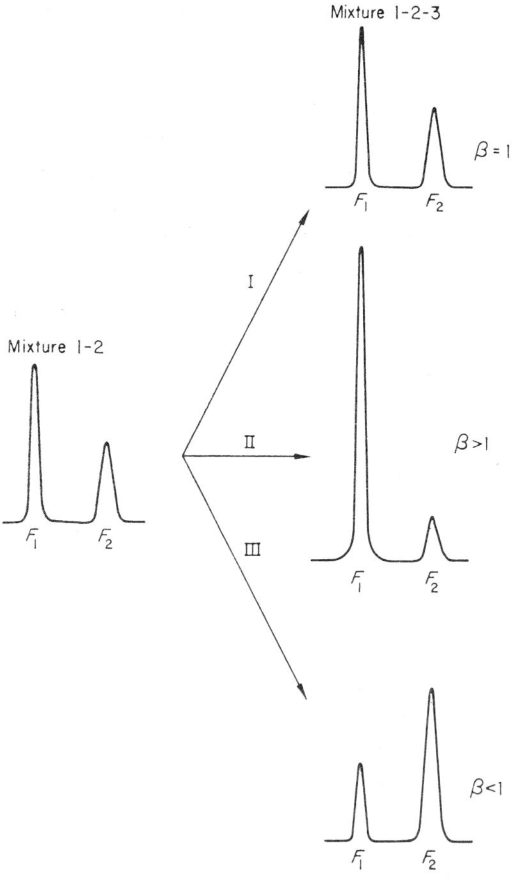

Fig. 72 Schematic representation of the three possible ways in which a solvent can change the relative peak areas.

In contrast to the well-established gas chromatographic method,[37-44] the solvent being evaluated is in the headspace analysis vessel (Fig. 71) rather than in the chromatographic column. However, if the solvent has a comparatively high boiling point such that it can be used as a stationary phase for gas chromatography, and if as the stationary phase it is able to resolve the two-component mixture of 1 and 2, then it is possible to combine the two techniques (Fig. 73) and evaluate the solvent simultaneously in the headspace analysis vessel and in the chromatographic column. In this case both the improvement factor β or β^* and also the selectivity S^* can be determined with the gas chromatograph by means of the equation

$$S^* = (p_{o2}/p_{o1})(t_2/t_1)_{12\infty 3} \tag{54}$$

in which t_1 and t_2 are the retention times measured for the compounds 1 and 2.

The above equation is derived from the relationship found by Röck:[37]

$$(\alpha_{12})_{12\infty 3} = (t_2/t_1)_{12\infty 3} \tag{55}$$

in which $(\alpha_{12})_{12\infty 3}$ is the relative volatility of compounds 1 and 2 at infinite dilution in compound 3, and $(t_2/t_1)_{12\infty 3}$ is the ratio of the reciprocal retention times of these compounds, based on the gas chromatographic separation of them using compound 3 as the stationary phase. Equation (54) is obtained from Eq. (55) by substitution of Eqs. (13b) and (34a).

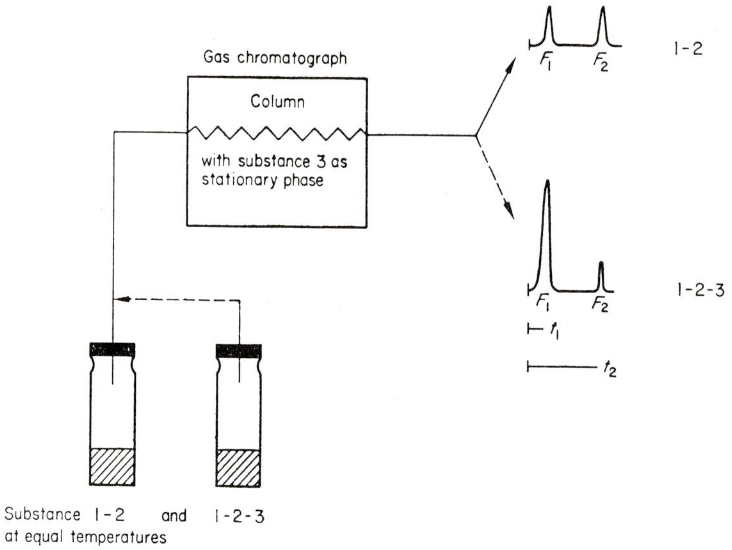

Fig. 73 Schematic representation of the combined gas chromatographic headspace analysis determination of the improvement factor β or β^* and the selectivity S^*.

In measurements performed by the combined method care must be taken to ensure that the temperatures of the headspace vessel and the gas chromatographic column are exactly the same. Since it is necessary to prepare and install a special column for each solvent, measurements can only be performed automatically if a series of determinations is to be carried out with a given solvent.

The selectivities can be determined without making use of the combined method if the pure compounds 1 and 2 are subjected to gas chromatographic headspace analysis under the same conditions as the two-component mixture 1–2 and the three-component mixture 1–2–3. In this case, application of Eqs. (34) or (34a) and (47a) to the two substances 1 and 2 gives the relationships

$$S = (F_{02}/F_{01})(x_2/x_1)_{123}(F_1/F_2)_{123} \tag{56}$$

and

$$S^* = (F_{02}/F_{01})(x_2/x_1)_{12\infty 3}(F_1/F_2)_{12\infty 3} \tag{56a}$$

which contain only measurable quantities, i.e. the mole fractions x_1 and x_2 in the three-component mixture, which are derived from the composition of the mixture prepared from measured volumes or masses, and the peak areas determined in the comparative gas chromatographic headspace analyses with the Multifract Automat (F40 and F42, Perkin-Elmer, Bodenseewerk). The peak areas F_{01} and F_{02} refer to the pure substances 1 and 2 as the sample, and F_1 and F_2 refer to their peak areas obtained from the analysis of the three-component mixture.

In the same way that suitable solvents can be found for extractive distillation by determining the selectivity S of two compounds 1 and 2 in different solvents, it is possible to select liquid stationary phases for gas chromatographic analysis by determining the selectivity at infinite dilution S^*, and applying Eqs. (54a) and (56a):

$$t_2/t_1 = (p_{01}/p_{02})S^* \tag{57}$$

$$t_2/t_1 = (p_{01}/p_{02})(F_{02}/F_{01})(x_2/x_1)_{12\infty 3}(F_1/F_2)_{12\infty 3} \tag{58}$$

These equations give the relative retention times t_2/t_1 providing, of course, that the vapour pressures of the pure substances p_{01} and p_{02} are known at the temperature of the measurements. This method has been recommended by Hachenberg.[73]

2.22. The Determination of Vapour–Liquid Equilibrium Data for the Characterization of Solvents in Terms of Partition Coefficients

In recent years solvent effects have received increasing attention,[60–67] on the one hand because in experimental investigations in widely different fields the influence of solvents on substances dissolved in them can no longer be simply overlooked owing to the improvements in experimental accuracy which have been obtained, and on the other hand, because the theory of solutions is not yet sufficiently well developed for the solvent effects which are observed to be quantitatively described.

The primary effect of a solvent on a substance dissolved in it is to change the chemical potential of the solute in the liquid phase. The result of this can be observed in all those measurements where the values obtained are a function of the chemical potential of the solute such as, for example, in measurements of phase equilibrium data, of the equilibria in chemical reactions, of equilibria involving activated complexes in chemical reactions (application of the transition state theory to the rate constants of chemical reactions), and in measurements of solvent shifts of solute molecules by spectroscopic methods (electron excitation-, infrared-, nuclear magnetic resonance and electron spin resonance spectra), of the relative volatilities of dissolved substances (selection of extraction agents for extractive distillation) and of relative retention times in gas chromatography.

In this connection Rohrschneider[68] has described a new method of determining gas–liquid partition coefficients and relative retention times by gas chromatographic headspace analysis, and has demonstrated the method with the Multifract F40 instrument (Perkin-Elmer, Bodenseewerk). The experimental results reported by Rohrschneider were the partition coefficients of *n*-octane, toluene, ethanol, methylethylketone, dioxane and nitromethane, obtained by comparative measurements with 80 different solvents.

The ratio between the partition coefficient* of toluene and that of n-octane was called the 'aromatic selectivity' and for each solvent it was compared with the molar volume of the solvent, the solubility parameter δ introduced by Hildebrand,[33] the E_T values employed by Reichardt and Dimroth[60] and derived from the band shift of 2,4,6-triphenyl-*N*-3,5-diphenyl-hydroxyphenyl-pyridinium-betaine, the solvent strength ε^0 of Snyder,[69] and finally with the solubility of three polymers.

In order to explain the theory of the method it is best to start from the equation defining the partition coefficient K:

$$K_i = C'_i/C''_i \tag{59}$$

where C'_i is the concentration of the solute i in the liquid phase, and C''_i is the concentration of i in the vapour phase, the concentrations being in mg cm^{-3} or mmol cm^{-3}.

The quantity m_i of the solute originally introduced into the pure solvent distributes itself between the vapour and the liquid phases, and the following equation can therefore be applied to the equilibrium:

$$m_i = V'C'_i + V''C''_i \tag{60}$$

in which V' is the volume (in cm^3) of the liquid phase and V'' is the volume (in cm^3) of the vapour phase.

$K'_iC''_i$ obtained from Eq. (59) may be substituted for C'_i, and rearranging the equation obtained gives the expression

$$K_i = (m_i/C''_i - V'')/V' \tag{61}$$

in which all the quantities on the right-hand side can be measured.

Instead of the concentration C''_i in the vapour phase Rohrschneider introduced the product of the measured peak height h_i and a proportionality factor r_i, which is specific for each substance:

$$C''_i = r_ih_i \tag{62}$$

$$K_i = m_i/(r_ih_iV') - V''/V' \tag{63}$$

The ratio m_i/r_i is independent of the nature of the solvent, and was first determined for a single selected substance (standard substance) i the partition

* This quantity is really a separation factor.

coefficient of which had been calculated by Khan[70] and Littlewood[71] from the expression

$$K_i = \frac{RT \cdot d_L}{p_{0i} \cdot \gamma_{i\infty L} \cdot M_L} \tag{64}$$

Data were taken from the literature for the density of the solvent d_L, the vapour pressure p_{0i} and the activity coefficient $\gamma_{i\infty L}$ of the substance i at infinite dilution in the solvent L. M_L is the molecular weight of the solvent. Rohrschneider then used the equation

$$m_i/r_i = V''/V' + K_i h_i V'$$

but this is incorrect, and should be written in the form

$$m_i/r_i = V''h_i + K_i h_i V' \tag{65}$$

which can easily be derived by transposing Eq. (63). Nevertheless, for the determination of partition coefficients this error has no effect, since obviously only one 'standard factor' m_i/r_i has been employed, (substance i = *n*-octane; solvent = squalane) and as a consequence of the high value of the partition coefficient K_i (=3512) the first term on the right-hand side of Eq. (65) can be neglected.

All of the other standard factors were calculated from the equation

$$m_j/r_j = (m_i/r_i)(h_j/h_i) \tag{66}$$

the ratio between the peak heights being determined by comparative gas chromatographic analysis. According to Rohrschneider, the errors in the partition coefficients which he determined should not exceed 15% with a 95% confidence limit. This also applies to his compilation of values of partition coefficients calculated from data in the literature by means of Eq. (64), and compared with those obtained from measurements with the Multifract F40 instrument.

Only about 3% of all the values measured were below 100.

The magnitude of the effect due to the change in the concentration in the liquid phase while equilibrium is being established can be easily demonstrated by means of Eq. (67), which is obtained from Eq. (59) by substituting n_i/V instead of C_i and rearranging it:

$$\frac{n'_i}{n''_i} = \frac{V'}{V''} K_i \tag{67}$$

If the ratio between the volume of the liquid phase and that of the vapour phase is taken to be the same as in Rohrschneider's measurements, i.e. $V'/V'' = 2.0/11.4$, and K_i is assumed to be 100, then the ratio between the number of moles in the liquid phase and that in the vapour phase is calculated to be $n'_i/n''_i = 17.5$, and the number of moles originally present in the liquid phase is: $n_i = n'_i + n''_i = 1.057n'_i$. While the equilibrium is being established, the concentration of substance i therefore decreases by only 5.7% of its initial value, in the unfavourable case where the partition coefficient is only 100. When the errors in the results are assessed this effect can therefore be neglected in those cases where the measured partition coefficient is not less than 200.

If n'_i in Eq. (67) is replaced by the difference $n_i - n''_i$ and the equation is transformed, it becomes

$$K_i = (n'/n''_i - 1)V''/V' \tag{68}$$

and substitution of C_i and C''_i instead of n_i/V' and n''_i/V'' gives

$$K_i = C_i/C''_i - V''/V' \tag{69}$$

In this expression for K_i the concentration C_i is known from the amounts of material used to prepare the solution, the concentration C''_i represents the quantity actually measured in the gas chromatographic headspace analysis, and V' and V'' are the volumes of the liquid phase and the vapour phase and can easily be determined. If the partition coefficient has a high value, the second term in Eq. (69) is negligible, but it must be taken into account when the partition coefficient itself is small.

The importance of the partition coefficient K_i is partly due to the fact that the ratio between the retention times of two substances 1 and 2 measured from the inert gas peak is equal to the ratio between the adjusted retention volumes V_1 and V_2, which is equal to the ratio between the corresponding partition coefficients K_1 and K_2:

$$t_1/t_2 = V_1/V_2 = K_1/K_2 \tag{70}$$

Moreover, a knowledge of the partition coefficient K_i is essential for the determination of absolute retention data, since, according to Kahn[70] and Littlewood,[71]

$$K_i = V_g{}^T d_L \tag{71}$$

where d_L again represents the density of the solvent (stationary phase) and $V_g{}^T$ is the adjusted retention volume, corrected for the pressure drop of the column, per gram of solvent at the temperature at which the partition equilibrium is established.

The combination of Eqs. (71) and (69) gives a relationship between the measured peak area F''_i for the material in the headspace and the retention volume $V_g{}^T$:

$$V_g{}^T d_L = C_i/C''_i - V''/V' \tag{72}$$

Equation (38) can be used to substitute the quantity $(F''/c_i)/V''$ instead of the concentration $C''_i = n''_i/V''$ in Eq. (72) to give

$$V_g{}^T d_L = C_i/[(F''_i/c_i)/V''] - V''/V' \tag{73}$$

and further rearrangement finally gives

$$V_g{}^T d_L = (C_i V'')/(F''_i/c_i) - V''/V' \tag{74}$$

The final term in this equation can be neglected if the partition coefficient K_i is not too small. If, during comparative measurements in which different solutes are used, the density d_L and also the concentration C_i and the volume V'' are kept constant, then normalized peak areas $\bar{F}''_i = F''_i/c_i$ are obtained which are inversely proportional to the retention volume V_g^T.

This inverse relationship between the retention volume and the normalized peak area was established experimentally by Hachenberg[73] from measurements of retention times t_i, which are proportional to the retention volumes, and of peak areas F''_i, which were subsequently normalized.

TABLE 13

Peak areas and retention times* obtained by comparative gas chromatographic headspace analysis of samples of glycerine containing various alcohols

Solute	*Solvent*	t (min)	F''_i	c	$\bar{F}''$
Isopropanol	Glycerine	10.4	1221	1.00	1221
Ethanol	Glycerine	14.0	769	1.43	1100
n-Propanol	Glycerine	15.1	616	0.95	585
Methanol	Glycerine	16.3	281	2.07	582
n-Butanol	Glycerine	18.0	563	1.02	574

* *The retention times were determined in an ordinary gas chromatograph using the following operating conditions:*
Instrument: model F7 chromatograph, Perkin-Elmer, Bodenseewerk
Detector: FID
Column packing: 25 *wt.%* *Glycerine on* 60/100 *mesh kieselguhr*
Column length: 6 m
Internal diameter of column: 4 mm
Sample size: 5 μl
Carrier gas: N_2
Column temperature: 70°C
Display: potentiometer recorder

The concentrations and response factors in Table 13 are based on the composition by weight rather than mole fractions. The concentration C_i was always 1 wt.% and the temperature of the thermostat bath in the F40 instrument was 70°C.

2.3. POTENTIAL PHYSICO-CHEMICAL APPLICATIONS OF HEADSPACE ANALYSIS

We have so far dealt exclusively with physico-chemical applications of gas chromatographic headspace analysis, the feasibility of which have already been established by detailed experimental studies. Finally, however, some reference should be made to other applications which might be developed in physical chemistry.

Mention should first be made of the fact that although it has been established that vapour pressures and heats of vaporization can be measured,[74] a large number of determinations still remain to be carried out. Moreover, the accuracy with which heats of mixing and heats of solution can be determined, and the amount of effort required to make the measurements is yet to be established.

In studies of chemical reactions it should be possible to determine equilibrium constants, reaction rate constants and the orders of reactions, etc., from measurements of concentrations and activities.

Finally, it is conceivable that physico-chemical properties such as the heats of adsorption, adsorption isotherms, specific surface areas and porosities of solid stationary phases could be determined by this technique.

Future developments will show whether gas chromatographic headspace analysis can provide useful, or even unexpectedly valuable, results in these and perhaps other branches of physical chemistry.

References

1. LEWIS, G. N. and RANDALL, M., *Thermodynamics*, McGraw-Hill, New York (1961).
2. VAN NESS, H. C., *Classical Thermodynamics of Non-Electrolyte Solutions*, Pergamon, Oxford (1964).
3. DENBIGH, K. G., *Prinzipien des Chemischen Gleichgewichts*, Dr. Dietrich Steinkopff Verlag, Darmstadt (1959).
4. KORTÜM, G., *Einführung in die Chemische Thermodynamik*, Vandenhoeck & Ruprecht in Göttingen Verlag Chemie GmbH, Weinheim/Bergstr (1960).
5. HAASE, R., *Thermodynamik der Mischphasen*, Springer-Verlag, Berlin (1956).
6. HALA, E., PICK, J., FRIED, V. and VILIM, O., *Vapour–liquid Equilibrium*, Pergamon, Oxford (1967).
7. PRAUSNITZ, J. M., ECKERT, C. A., OREY, R. V. and O'CONNELL, J. P., *Computer Calculations for Multicomponent Vapor–Liquid Equilibria*, Prentice-Hall, Englewood Cliffs, N.J. (1967).

8. PRAUSNITZ, J. M., *Molecular Thermodynamics of Fluid-Phase Equilibria*, Prentice-Hall, Englewood Cliffs, N.J. (1969).
9. ONKEN, U., *Thermische Verfahrenstechnik*, Carl Hanser Verlag, Munich (1975).
10. KOGAN, W. B. and FRIDMAN, W. M., *Handbuch der Dampf-Flüssigkeits-Gleichgewichte*, VEB Deutscher Verlag der Wissenschaften, Berlin (1961).
11. D'ANS, J. and LAX, E., *Taschenbuch für Chemiker und Physiker*, Vol. I, Springer-Verlag, Berlin (1967).
12. WICHTERLE, I., LINEK, J. and HALA, E., *Vapour–Liquid Equilibrium*, Data Bibliography, Elsevier, Amsterdam (1973).
13. LANDOLT-BÖRNSTEIN, *Zahlenwerte und Funktionen aus Naturwissenschaft und Technik*, New Series, Vol. 3, *Thermodynamisches Gleichgewicht siedender Gemische*, Springer-Verlag, Berlin (1975).
14. KOLB, B., *CZ—Chem.-Technik*, **1**, 87 (1972).
15. WICHTERLE, I. and HALA, E., *Ind. Eng. Chem. Fundamentals*, **2**, 155 (1963).
16. DIETRICH, H. J., *Das Dampf-Flüssigkeits-Gleichgewicht des Testgemisches Benzol*/1,2-*Dichloräthan im Bereich von* 10 *bis* 80°*C*, Dissertation Universität Stuttgart (1974).
17. MONTFORT, J. P., 'Experimental determination of the Wilson equation', *5th Int. Congress on Chemical Engineering, Chemical Equipment Design and Automation* (CHISA), Prague, 25–29 August (1975).
18. KORTÜM, G. and BUCHHOLZ-MEISENHEIMER, H., *Die Theorie der Destillation und Extraktion von Flüssigkeiten*, Springer-Verlag, Berlin (1952).
19. REDLICH, O. and KWONG, J. N. S., *Chem. Rev.*, **44**, 233 (1949).
20. WOHL, K., *Trans. Am. Inst. Chem. Engng.*, **42** (1946).
21. OREY, R. V. and PRAUSNITZ, J. M., *Ind. Engng. Chem.*, **57**, 18 (1965).
22. RENON, H. and PRAUSNITZ, J. M., *AIChE J.*, **14**, 135 (1968).
23. VAN WINKLE, M., *Distillation*, McGraw-Hill, New York (1967).
24. ROWLINSON, J. S., *Liquids and Liquid Mixtures*, Butterworth, London (1959).
25. KRELL, E., *Handbuch der Laboratoriumsdestillation*, VEB Deutscher Verlag der Wissenschaften, Berlin (1960).
26. SURYANARAYANA, Y. S. and VAN WINKLE, M., *J. Chem. Engng. Data*, **11**, 7 (1966).
27. KUMAR, R., PRAUSNITZ, J. M. and KING, C. J., *Advan. Chem. Ser.*, No. 115, 16 (1972).
28. BITTRICH, J., *Z. Chem.*, **10**, 201 (1970).
29. TASSIOS, D., *Chem. Engng.*, 118 (1969).
30. HÜWELS, W., *CZ-Chem.-Technik*, **3**, 85 (1974).
31. STEIN, M., *Hydrocarbon Processing*, 139 (1973).
32. PIEROTTI, G. J., DEAL, C. A. and DERR, E. L., *Ind. Engng. Chem.*, **41**, 95 (1959).
33. HILDEBRAND, J. H. and SCOTT, R. L., *The Solubility of Non-Electrolytes*, Reinhold, New York (1950).
34. WEIMER, R. F. and PRAUSNITZ, J. M., *Hydrocarbon Processing*, **44**, 237 (1965).
35. HELPINSTILL, J. G. and VAN WINKLE, M., *I & EC Process Design and Development*, **7**, 213 (1968).
36. NULL, H. R. and PALMER, D. A., *Chem. Engng. Progr.*, **69**, 47 (1969).
37. RÖCK, H., *Chem. Ing. Techn.*, **28**, 489 (1956).
38. KELKER, H., *Ullmanns Encyclopädie der Technischen Chemie*, 3rd edn, Vol. II/1, 149 (1961).
39. WARREN, G. W., WARREN, R. R. and YARBOROUGH, V. A., *Ind. Engng. Chem.*, **51**, 1475 (1959).
40. DÖRING, C., *Z. Chem.*, **1**, 347 (1961).
41. PORTER, R. S. and JOHNSON, J. F., *Ind. Engng. Chem.*, **52**, 691 (1960).

42. PIEROTTI, G. J., DEAL, C. H. and PORTER, P. E., *J. Am. Chem. Soc.*, **78**, 2989 (1956).
43. PORTER, R. E., DEAL, C. H. and STROSS, F. H., *J. Am. Chem. Soc.*, **78**, 2999 (1956).
44. PORTER, R. S. and JOHNSON, J. F., *Nature*, **183**, 391 (1959).
45. GARNER, F. H. and ELLIS, S. R. M., *Trans. Ind. Chem. Engrs.* (*London*), **29**, 45 (1951).
46. QUOZATI, A. and VAN WINKLE, M., *J. Chem. Engng. Data*, **5**, 269 (1960).
47. CHEARS, M. N. and MAKIN, E. C., *Am. Chem. Soc. Paper Regional Meeting St. Louis* (December 1–3, 1960).
48. KOLB, B., *J. Chromatog.*, **112**, 287 (1975).
49. BEATTIE, J. A., *Chem. Rev.*, **44**, 141 (1949).
50. BEATTIE, J. A. and STOCKMAYER, W. H., *Treatise on Physical Chemistry*, Van Nostrand, Princeton, N.J. (1942).
51. BEATTIE, J. A., *Thermodynamics and Physics of Matter*, Princeton University Press (1955).
52. GREWER, TH. and SCHMIDT, A. P., *Chem. Ing. Techn.*, **45**, 1063 (1973).
53. HACHENBERG, H., unpublished work.
54. KETCHUM, R.-G., ROSCHER, G. and SCHMIDT, A. P., *Verfahrenstechnik* **10**, 116 (1976).
55. SCHMIDT, A. P., unpublished results.
56. RÖCK, H. and SIEG, L., *Z. Phys. Chem.*, **3**, 355 (1955).
57. BOUBLIK, T., FRIED, V. and HALA, E., *The Vapour Pressures of Pure Substances*, Elsevier, Amsterdam (1973).
58. HACHENBERG, H. and SCHMIDT, A. P., *Verfahrenstechnik*, **8**, 343 (1974).
59. HECK, G. and SCHMIDT, A. P., *Chem. Ing. Techn.*, **47**, 541 (1975).
60. REICHARDT, CH. and DIMROTH, K., *Fortschr. Chem. Forsch.*, **11**, 1 (1968/69).
61. ROHRSCHNEIDER, L., *Fortschr. Chem. Forsch.*, **11**, 146 (1968/69).
62. FUCHS, O., *Fortschr. Chem. Forsch.*, **11**, 74 (1968/69).
63. HEYDTMANN, H., SCHMIDT, A. P. and HARTMANN, H., *Ber. Bunsenges. Phys. Chem.*, **70**, 444 (1966).
64. DROUGARD, Y. and DECROOCQ, D., *Bull. Soc. Chim. France*, **1969**, 2972.
65. BERREBI, G. and DECROOCQ, D., *J. Chim. Phys.*, **71**, 673 (1974).
66. ECKERT, C. A., *Ind. Eng.*, **59** (9), 20 (1967).
67. ECKERT, C. A., HSIEH, C. K. and McCABE, J. R., *AIChE J.*, **20** (1), 20 (1974).
68. ROHRSCHNEIDER, L., *Anal. Chem.*, **45**, 1241 (1973).
69. SNYDER, L. R., *Principles of Adsorption Chromatography*, Marcel Dekker, New York (1968).
70. KHAN, M. A., *Lab. Pract.*, **11**, 195 (1962).
71. LITTLEWOOD, A. B., *Gas Chromatography*, Academic Press, New York (1962).
72. HILDEBRAND, J. H. and SCOTT, R. L., *The Solubility of Nonelectrolytes*, p. 181, Dover, New York (1964).
73. HACHENBERG, H., unpublished work.
74. VAN STEKELENBURG, G. J. and KOOREVAAR, G., *Clin. Chim. Acta*, **34**, 305 (1971).
75. DYMOND, J. H. and SMITH, E. B., *The Virial Coefficients of Gases*, Clarendon Press, Oxford (1969).

SUBJECT INDEX

Q

R

S

T

U

V

W

CHEMICAL COMPOUND INDEX

O

P

S

T

V

W

X